T0340066

Auditory Contributions to Food Perception and Consumer Behaviour

Auditory Contributions to Food Perception and Consumer Behaviour

Edited by

Charles Spence, Felipe Reinoso-Carvalho,
Carlos Velasco and Qian Janice Wang

BRILL

Leiden • Boston
2019

Originally published as Volume 32, NO (4-5) 2019 of Brill's journal *Multisensory Research* (MSR).

Library of Congress Cataloging-in-Publication Data

Names: Spence, Charles (Experimental psychologist), editor. | Reinoso Carvalho, Felipe, editor. | Velasco, Carlos (Professor of marketing), editor. | Wang, Qian Janice, editor.
Title: Auditory contributions to food perception and consumer behaviour / edited by Charles Spence, Felipe Reinoso-Carvalho, Carlos Velasco and Qian Janice Wang.
Description: Leiden ; Boston : Brill, 2019. | Includes bibliographical references and index.
Identifiers: LCCN 2019030335 (print) | LCCN 2019030336 (ebook) | ISBN 9789004416284 (hardcover) | ISBN 9789004416307 (ebook)
Subjects: LCSH: Food–Sensory evaluation. | Sound–Psychological aspects. | Consumer behavior.
Classification: LCC TX546 .A93 2019 (print) | LCC TX546 (ebook) | DDC 664/.072–dc23
LC record available at https://lccn.loc.gov/2019030335
LC ebook record available at https://lccn.loc.gov/2019030336

ISBN: 978-90-04-41628-4 (hardback)
ISBN: 978-90-04-41630-7 (e-book)

CONTENTS

Note on Contributors

Sue Bastian is Associate Professor in Oenology and Sensory Studies, at the University of Adelaide with over two decades of experience working on beverage and food composition, sensory, quality and consumer research. Sue is an oenologist, protein biochemist, and cell biologist. Her research examines the context, inter-individual, sensory and molecular drivers of consumer food choice and preference. Current research focuses on cross-cultural market-insights, virtual reality, and emotions; genetics of mouthfeel perception; terroir of Australian Shiraz wine; wine tourism; authenticity; wine and food pairing; crossmodal interactions; improving sensory and consumer methodologies and traditional Chinese medicine and plant extracts in cancer and gut health. She has published nearly 100 peer-reviewed scientific articles.

Thadeus L. Beekman is currently a Ph.D. student in the Department of Food Science, University of Arkansas, Fayetteville, AR, USA. Beekman grew up in Saline, MI, USA and received his B.S. degree from the Department of Food Science and Human Nutrition at the Iowa State University, Ames, IA, USA. His research interests include investigating the factors influencing the way individuals think about and react during sensory testing. He is also interested in investigating the impact of environmental conditions on consumers' multisensory perception and behavior.

Jo Burzynska is a sonic artist, writer and curator with a two-decade practice that spans experimental music performance and recording to public and gallery installation art. Jo is also a widely-published wine writer, international judge and author of *Wine Class* (Penguin Random House, 2009), her work in both areas has increasingly converged in the production of artworks that combine sound and taste. She is currently researching perceptual and aesthetic interactions between sound and wine and their application in a multisensory arts practice through a Ph.D. at the University of New South Wales, Sydney (2016–2020).

Andrew Childress received his B.A. in English Literature and B.M. in Music (Piano Performance) from the Departments of English and Music, respectively, at the University of Arkansas, Fayetteville, AR, USA. He received a Master of Science in Information Studies from the University of Texas at

© KONINKLIJKE BRILL NV, LEIDEN, 2019 | DOI:10.1163/9789004416307_002

Austin, Texas, USA. He is currently working as a Senior Taxonomy Analyst at Indeed in Austin.

Ilja Croijmans is a psychologist in the area of olfaction, currently working at Utrecht University, the Netherlands, where he investigates chemosensory communication in humans. He received his PhD from Radboud University, the Netherlands, with his dissertation titled *Wine expertise shapes olfactory language and cognition* in 2018. In his dissertation, he investigated the influence of olfactory expertise on cognition, including language, memory, and imagery.

Silvana Dakduk is a social psychologist, with a Ph.D. degree in psychology from the Catholic University Andrés Bello (Venezuela). She has over seventeen years of academic experience in different countries across Latin America. Her research interests are related to consumer behavior and pedagogical innovation in higher education, as well as in executive education. At the moment, she is a professor of Marketing at Universidad de los Andes — School of Management, in Colombia.

Alexandra Fiegel is a sensory and consumer scientist. Fiegel received her B.S. from the Department of Biology, University of Iowa, Iowa City, IA, USA. She also holds a M.S. in Sensory and Consumer Science from the Department of Food Science, University of Arkansas, Fayetteville, AR, USA. Her thesis title is "Influences of music genre and components on food perception and acceptance". She is currently working as an Associate Sensory Scientist at Cargill, Wayzata, MN, USA.

Apratim Guha is a professor of statistics at Xavier Labour Research Institute, India. His research focuses on applied statistics and information theory. He is particularly interested in applications of statistical methods in biological sciences, computer science and management. He obtained his PhD from the department of statistics, University of California, Berkeley, and has taught previously at National University of Singapore, University of Birmingham, UK, and Indian Institute of Management, Ahmedabad, India.

Ryuta Kawashima is a Director of the Smart Ageing International Research Center, Institute of Development, Aging and Cancer, Tohoku University. He has won the Prizes for Science and Technology, The Commendation for Science and Technology by the Minister of Education, Culture, Sports, Science and Technology. His scientific output includes over 400 peer-reviewed papers and numerous books. His scientific interest is in functional brain mapping of higher cognitive functions of humans. He has also become increasingly interested in returning benefits of basic sciences to the public, so that he has proposed systems for improvement of the cognitive functions of humans by industry-university cooperation.

Bruno Mesz is a musician, multisensory artist, and mathematician. As a researcher, he works on sound perception, musical semantics, multisensory perception and mathematical models of music at Universidad Nacional de Tres de Febrero, Buenos Aires, Argentina, where he is a professor at the Arts and Culture Department. He has presented multisensory works and installations at the Cooper Hewitt Design Museum (New York), Reid Hall (Paris), Seinajoki Tango Festival in Finland, MALBA (Buenos Aires), among others. He is a pianist and clarinetist, specialized in contemporary music.

Kosuke Motoki is Assistant Professor at Department of Food Management, School of Food, Agricultural and Environmental Sciences, Miyagi University, Japan. His research focuses on sensory and consumer science, particularly contextual influences on consumer responses to food products. He studies how sensory, emotional, and social factors contribute to consumer perception and preferences, using behavioral experiments as well as neurophysiological techniques (e.g., eye-tracking and fMRI). His research has been published in *Food Quality and Preference*, *Appetite*, *Multisensory Research*, and *Frontiers in Psychology*.

Rui Nouchi is Associate Professor at Department of Cognitive Health Science, Institute of Development, Aging and Cancer, Tohoku University, Japan. His research interest is to understand and improve cognitive health for an individual and a society. He coined the new phrase "Cognitive Health Science", which is a research area investigating healthy and adaptive mental states (cognitive and emotional) throughout the lifespan. To understand and improve Cognitive Health, he uses multiple research methods such as psychological experiments and questionnaire methods, brain imaging methods, daily intervention methods using randomized controlled trials, and meta-analysis methods.

Felipe Reinoso-Carvalho is an assistant professor of marketing and consumer behavior, at Universidad de los Andes School of Management — a triple crown business school located in Bogotá, Colombia. He holds a joint-PhD degree in engineering sciences (Vrije Universiteit Brussels), and psychology (University of Leuven/KU Leuven). In his research, he focuses on designing meaningful experiences for consumers, by blending the rationale of experiential marketing, and multisensory perception. He advocates for the use of science, and for an appropriate understanding of technology, as a means for achieving a better future. In parallel of his tenure position at Universidad de los Andes, he holds invited scholar positions at KU Leuven and IAE Angers.

Pablo Riera is a researcher and musician. He holds a Ph.D. in Physics from the University of Buenos Aires and teaches signal processing for music analysis at the University of Quilmes. He currently works at the Applied Artificial Intelligence Lab at UBA where he researches on machine learning for sound processing. As a musician, he has performed live shows using EEG, Neural Networks and differential equations.

Marijn Peters Rit formerly studied Communication and Information Sciences at the Universities of Nijmegen and Tilburg. During her master's in Nijmegen at Radboud University, her research focused on marketing communication. She explored the more technical aspects of communication during her master's at Tilburg University, where she investigated human–robot communication.

Toshiki Saito is a JSPS Research Fellow and Ph.D. candidate at Tohoku University School of Medicine. Before entering the Ph.D. course at Tohoku University, he obtained a master's degree in Psychology at Meiji Gakuin University, Japan. His research focuses on social cognition. He investigates how people perceive and process other's information using behavioral experiments, eye-tracking techniques, and neuroscientific methods. His research has been published in *Frontiers in Aging Neuroscience*, and *Cognitive Studies*.

Han-Seok Seo is currently an Associate Professor and Director of the University of Arkansas Sensory Service Center in the Department of Food Science at the University of Arkansas, Fayetteville, AR, USA. He received a Ph.D. in Food and Nutrition and a Doctor of Medical Science in Otorhinolaryngology from Seoul National University (Seoul, Republic of Korea) and the Technical University of Dresden (Dresden, Germany), respectively. He serves as an editorial board member for multiple journals, including the Journal of Sensory Studies, Food Quality and Preference, Foods, and Journal of Culinary Science & Technology.

Mariano Sigman is the director of the Neuroscience Laboratory at Di Tella University (Buenos Aires), an interdisciplinary group integrated by physicists, psychologists, biologists, engineers, educational scientists, linguists, mathematicians, artists and computer scientists. He studied physics at the University of Buenos Aires, his PhD in neuroscience in New York, and a postdoc at College de France, Paris. He has published more than 150 articles in the most prestigious neuroscience, physics, economy, and psychology journals. He investigates decision-making and confidence, and how current knowledge of the brain and the mind may serve to improve educational practice. He also has a research program combining basic science with practical questions of neurology and psychiatry, developing tools to improve diagnosis in schizophrenia, depression, and in comatose patients. Mariano was awarded a Human Frontiers Career Development Award, the national prize of physics, the young investigator prize of "College de France", the IBM Scalable Data Analytics award The Pius XI Medal of the Pontifical Academy of Sciences and is a scholar of the James S. McDonnell Foundation.

Laura J. Speed is an experimental psychologist working in the Department of Psychology at the University of York, UK. She completed her PhD in Cognitive, Perceptual, and Brain Sciences at University College London. Her main area of research is the interaction between language and perception, focusing on how the perceptual systems are recruited during language comprehension, and how language affects perception.

Charles Spence is a professor of experimental psychology at Oxford University. His prize-wining research at the Crossmodal Research Laboratory investigates the factors that influence what we eat and what we think about the experience with world-leading chefs and food and beverage companies. He is the author of the Prose prize-winning "The perfect meal" with Betina Piqueras-Fiszman (2014, Wiley-Blackwell), the international bestseller "Gastrophysics: The new science of eating" (2017; Penguin Viking) — winner of the 2019 Le Grand Prix de la Culture Gastronomique from Académie Internationale de la Gastronomie, and Multisensory Packaging Design (2019; Palgrave MacMillan). He has published more than 900 peer-reviewed scientific articles. He is a regular on TV and Radio, and has been profiled by *The New Yorker* (http://www.newyorker.com/magazine/2015/11/02/accounting-for-taste; see also Charles Spence — *Sensploration, FoST 2016*; https://vimeo.com/170509976).

Motoaki Sugiura is a Professor at the Department of Human Brain Science, Institute of Development, Aging and Cancer, Tohoku University, Japan. To reveal the neural mechanisms and computational processes underlying the 'human' mind and behavior, he and his lab members from various academic fields are working together, combining functional neuroimaging technique with a variety of research methods. Through these research activities, he strives for a brain science that is considered a 'hub', connecting different fields of basic and applied sciences addressing human nature.

Marcos Trevisan is Professor of Physics at the University of Buenos Aires and researcher at the National Scientific and Technical Research Council (CONICET) in Argentina. His work deals with the physics and neuroscience of voice, and also with the evolution of language from the physical mechanisms of voice.

Carlos Velasco is an Associate Professor at the Department of Marketing, BI Norwegian Business School (Norway), where he co-founded the Center for Multisensory Marketing. He also holds a Research Fellowship at the SCHI Lab, Sussex University (UK). Carlos received his D.Phil. in Experimental Psychology from Oxford University. His work lies at the intersection between Psychology, Marketing, and Human–Computer Interaction, and focuses on understanding, and capitalizing on, our multisensory experiences and their guiding principles. Carlos is one of the founding members of Neurosketch (Colombia). He has worked with several companies from all around the world on topics such as multisensory experience design, food and drink, packaging, branding, and consumer research. For more information, see https://carlosvelasco.info/.

Johan Wagemans is a full professor in experimental psychology and currently director of the Department of Brain & Cognition at KU Leuven. He has a BA in psychology and philosophy, an MSc and a PhD in psychology (KU Leuven). He was a post-doc from the National Fund for Scientific Research, which allowed him to spend one year abroad in the laboratory of Michael Kubovy (U. Virgina). He is interested in perceptual organization (e.g., grouping, figure-ground, texture, depth, shape, objects), where he wants to contribute to the understanding of why things look as they do, at different levels (e.g., phenomenological, psychophysical, computational, neural, ecological). He is also interested in applications of our insights in autism, arts, and sports. He is supervising a long-term research program (2008–2022) aimed at reintegrating Gestalt psychology into contemporary vision science/neuroscience (see http://www.gestaltrevision.be). Recently, Prof. Wagemans edited The Oxford Handbook of Perceptual Organization (2015; Oxford University Press).

Qian Janice Wang is an assistant professor at the Department of Food Science at Aarhus University, in the Food Quality, Perception and Society group. Her research focuses on people's relationship with food, with on a special focus on how food-extrinsic factors, such as sound, can modify and enhance the way we perceive food and drink. Her work has been covered in publications such as the Financial Times, the Economist, La Revue du Vin de France, Le Figaro, and National Post Canada. She obtained her PhD from the Department of Experimental Psychology from the University of Oxford. Her work spans psychological experiments, technological enhancements, and multisensory performance.

Introduction to Auditory Contributions to Food Perception and Consumer Behaviour

Charles Spence [1,*], **Felipe Reinoso-Carvalho** [2], **Carlos Velasco** [3] and
Qian Janice Wang [4]

[1] Department of Experimental Psychology, University of Oxford, UK
[2] Universidad de los Andes School of Management, Bogota, Colombia
[3] Department of Marketing, BI Norwegian Business School, Norway
[4] Department of Food Science, Aarhus University, Denmark

1. Introduction

What we hear before and/or while we eat and drink can affect our taste/flavour experiences, even if we do not realize it. While there has been a long history of sensory science research investigating the impact of food sounds during consumption (e.g., Demattè *et al.*, 2014; Zampini and Spence, 2004; see Spence, 2015, for a review), the focus of this special issue is rather on the product-extrinsic, or contextual/situational, cues such as music and ambient soundscapes. Intriguingly, such auditory inputs have been shown to influence what we expect, what we choose to buy/order in shops, restaurants, cafes, and bars, and how satisfied we are with the experience, as well as the perception we have of what we eat and drink (e.g., Biswas *et al.*, 2018; Feinstein *et al.*, 2002; Lin, Hamid, Shepherd, Kantono, and Spence, 2019; Novak *et al.*, 2010; Septianto, 2016; Spence, 2012, 2017b; Zellner *et al.*, 2017).

One area of particular interest for the present volume relates to the recent emergence of a body of literature on the topic of 'sonic seasoning' (Spence, 2017a): This is where music/soundscapes are especially chosen, or else designed/composed, in order to correspond to, and hence hopefully to modify the associated taste/aroma/mouthfeel/flavour in food and beverages (e.g., see Crisinel *et al.*, 2012; Reinoso Carvalho *et al.*, 2015; Wang and Spence, 2016, for representative examples). Sonic seasoning, then, represents one strand of research on the crossmodal correspondences (see Spence, 2011, for a review).

---

[*] To whom correspondence should be addressed. E-mail: charles.spence@psy.ox.ac.uk

© KONINKLIJKE BRILL NV, LEIDEN, 2019 | DOI:10.1163/9789004416307_003

Interesting questions here concern where, exactly, such surprising crossmodal correspondences originate from, and the conditions under which they influence tasting experiences and eating/drinking behaviours — e.g., as a function of congruency (whether music and flavour correspond or not; e.g., Muniz *et al.*, 2017), or when compared to eating in silence (e.g., Hauck and Hecht, 2019; Höchenberger and Ohla, 2019; Lowe *et al.*, 2019; Watson and Gunter, 2017). Intriguingly, cognitive neuroscientists are slowly starting to turn their attention to the neural consequences/underpinnings of such 'almost-synaesthetic' crossmodal interactions (Callan, Callan and Ando, 2018).

One branch of the literature on sonic influences on tasting has also examined the phenomenon of 'sensation transference' (see Note 1) — addressing questions such as '*If you like the music more, do you like what you are eating/drinking more too?*' (e.g., see Kantono *et al.*, 2015, 2016; Wang and Spence, 2018; cf. Fritz *et al.*, 2017). A growing body of rigorous evidence now clearly suggests that it does. Not only can music and soundscapes affect the tasting experience, but a growing body of research shows that loud background noise (see Note 2), be it airplane noise, white noise, or restaurant noise, can influence both the taste of food and drink (e.g., higher levels of noise reduce sweetness perception; Rahne *et al.*, 2018; Woods *et al.*, 2011; Yan and Dando, 2015), as well as people's ability to discriminate various aspects of their tasting experience (see Spence, 2014a, for a review). Given the increasing presence of noise in many restaurants, cafes, and bars these days, this area of research may well have some important public health implications too: Just take the finding that noise levels correlate with both greater rates of consumption (e.g., Guéguen *et al.*, 2004, 2008; Kaiser *et al.*, 2016), and a tendency to make unhealthy food choices (Biswas *et al.*, 2018). Indeed, given the observation that what we hear can influence what we taste, one may consider sound as a relevant design element of food and beverage experiences — This is what chef Heston Blumenthal presumably had in mind when once he was quoted as saying '*Sound is the forgotten flavour sense*' (see Spence *et al.*, 2011). A growing body of evidence supports the view that many restaurants and bars in the west are becoming increasingly loud (Anon., 2019), and that such noise levels may be having a detrimental effect on both our long-term health and well-being as well as on the taste of the food (see Gourévitch *et al.*, 2014; Richard, 2012).

Here, it is important to note that the auditory inputs influencing our perception of what we eat/drink/taste are not limited to environmental sounds. They also involve the sounds that we produce while eating, such as slurping, crunching, or lip-smacking (Youssef *et al.*, 2017; see Spence, 2015, for a review), as well as speech sounds that we use to refer to specific foods (e.g., Spence, 2014b). The sounds that are associated with food containers, such as product packaging, are also relevant to tasting and decision-making (e.g., see

Spence and Wang, 2015; Wang and Spence, 2019). In many cases, such containers are a key part of the food experience itself, presumably through their role in setting expectations around the tasting experiences that are to come.

Given the growing literature on music and soundscape's influence on the multisensory tasting experience, there is a parallel interest in using technology to synchronize auditory stimulation with the tasting experience (see Velasco *et al.*, 2016; Spence, 2019a, for a review). Product extrinsic auditory–taste interactions are undoubtedly a rich area for creative practice, as evidenced by experiential events such as The Chocolate Symphony presented by Maxime Goulet (see http://symphonicchocolates.com/; www.maximegoulet.com). A growing number of experimental multisensory tasting events have been taking place in restaurants, bars, museums, and drinks events (e.g., Blecken, 2017; Royce-Greensill, 2017; see Spence, 2017b, 2019a, for a number of other such experiential multisensory tasting events). Both British Airways (with their Sound Bite menu for long-haul; see Victor, 2014), and, more recently, Finn Air with their long-haul flights to Asia, have also introduced an element of soundscape design to the meal experience (see https://www.finnair.com/cn/gb/stevenliu/en). Others, meanwhile, have been combining tasting with cultural events such as classical music concerts (Spence *et al.*, 2013). And, going one stage further, at the Tate Sensorium in London, visitors tasting bespoke chocolates while listening to a specially composed soundscape and viewing at one of Sir Francis Bacon's works (Davies, 2015; Pursey and Lomas, 2018). Marketing too are interested (e.g., see https://www.jagermeister.com/en/taste-remastered; https://www.youtube.com/watch?v=wvH_fYCoPzs; https://laughingsquid.com/the-seven-worlds-hennessy-x-o-cognac-ridley-scott/).

Despite its inauspicious beginnings 70 years ago (see Pettit, 1958), research specifically on atmospheric auditory contributions to food perception and behaviour has exploded in recent years, with interest coming from the fields of cognitive neuroscience, marketing, food science, gastronomy, design, branding, public health and beyond. In particular, in the present volume (based on a Special Issue that appeared initially in the journal *Multisensory Research*), we have brought together a series of papers on the influence of auditory cues on food perception and consumer behaviour. Taken together, these papers offer a snapshot of the state of the art of multisensory basic and applied research on the topic. Below, we present a short overview of the seven contributions in this volume.

2. Outline of Chapters

In the opening chapter, Spence *et al.* provide an overview of the literature concerning how food-unrelated extrinsic auditory cues can affect what we taste. In particular, the review covers key themes such as the influence of background noise and background music in food and drink experiences, as well as the role of sensation transfer and crossmodal correspondences. Next up, Reinoso Carvalho *et al.* provide evidence showing how affective transfer influences the perceived taste, and pleasantness of beers. These researchers report a series of experiments in which such crossmodal effects were analyzed when two contrasting music tracks (one associated with positive emotion, the other with negative emotion) were presented to consumers while tasting beer. The results suggest that the emotional reactions triggered by the music influenced specific sensory and hedonic aspects of the multisensory tasting experience (see also Kantono, Hamid, Shepherd, Lin, Skiredj, and Carr, 2019).

Speed, Peters, and Croijmans present a study designed to assess the effect of sound-odor congruency on participants' attitudes and memory for familiar and unfamiliar brands. Their results suggest that the level of arousal evoked by both auditory and olfactory cues is important when it comes to brand recognition. In particular, when participants were presented with high arousal sounds and smells together, their participants recognized the brands faster, relative to the other experimental conditions. This research adds to the literature on how affective congruence between sensory cues can influence the way in which consumers interact with brands. Motoki and colleagues continue the marketing theme, reporting on a series of studies demonstrating how the growing knowledge concerning crossmodal correspondences between audition and gustation are starting to be used in a marketing context in order to change what consumers choose/think about the food products that they choose.

In the next contribution, Fiegel *et al.* consider the influence of various parameters of auditory stimuli, including pitch, tempo, and volume levels of music in order to determine how they affect people's sensory perception and acceptance of foods (chocolate and bell peppers, one a hedonic food, the other not). Overall, the results revealed that the participants liked the foods more when music stimuli were perceived as more pleasant and stimulating, and that the flavors of the foods were perceived as more intense when the participants were more stimulated by the music samples that they heard. It is, though, perhaps worth bearing in mind here that a number of important questions remain here concerning the 'relative' *versus* 'absolute' nature of those crossmodal correspondences that involve the auditory dimension of pitch (see Spence, 2019b).

It is important to note that multisensory flavour perception is typically not a unitary and static experience. Rather, it is often complex and time-evolving. Recently-popular techniques from sensory science such as Rate-All-That Apply (RATA) and Temporal Dominance of Sensations (TDS) are increasingly enabling researchers to assess the various sensations that are experienced, for example, while savoring a mouthful of fine wine. In the final two articles in this volume, Burzynska and colleagues, and Wang *et al.* used such sensory methods in order to investigate the influence of sound on the perception of wine. Burzynska *et al.* evaluated the influence of high and low pitch on the perceived body, acidity, and intensity of two red wines whose flavour profiles were validated using the RATA approach. Meanwhile, Wang *et al.* used TDS to show that the temporal changes in flavours perceived in a mouthful of wine can be synchronised to changes in the background music.

3. Conclusions

In conclusion, the emerging multisensory connections that are being uncovered between sound and food constitutes an exciting field with the potential to encourage collaboration across multiple disciplines, to enhance our theoretical understanding of multisensory flavour perception, broadly-defined, as well as potentially to create applications with wider societal impact. With time, as the empirical literature and interest in this topic continues to grow, the underlying mechanisms that explain such surprising crossmodal associations and influences are starting to be disentangled. New methods and techniques are being successfully adopted, in the process bringing novel insights to the field, and as we have seen, marketing and management are starting to perk their ears up as they start to 'see' the relevance of these topics, with an explosion of multisensory experiential tasting events occurring in both the worlds of branding and popular culture (see Spence, 2017b, 2019a, for a number of examples).

Notes

1. This phenomenon, note, appears elsewhere in the multisensory literature under the label 'affective ventriloquism' (see Spence and Gallace, 2011).

2. When what we hear becomes too loud, it is usually regarded as noise.

References

Anon. (2019). Is Nigella right — can loud music really affect the taste of your food?, *The Daily Telegraph Online*, **June 6th**, 1, https://www.telegraph.co.uk/food-and-drink/features/nigella-right-can-loud-music-really-affect-perception-food/.

Biswas, D., Lund, K. and Szocs, C. (2018). Sounds like a healthy retail atmospheric strategy: effects of ambient music and background noise on food sales, *J. Acad. Mark. Sci.* **47**, 37–55.

Blecken, D. (2017). Hold the sugar: a Chinese café brand is offering audio sweeteners, *Campaign*, **February 13th**. https://www.campaignasia.com/video/hold-the-sugar-a-chinesecafe-brand-is-offering-audio-sweeteners/433757.

Callan, A., Callan, D. and Ando, H. (2018). Differential effects of music and pictures on taste perception — an fMRI study. Poster presented at the *Annual Meeting of the International Multisensory Research Forum*, Toronto, ON, Canada.

Crisinel, A.-S., Cosser, S., King, S., Jones, R., Petrie, J. and Spence, C. (2012). A bittersweet symphony: systematically modulating the taste of food by changing the sonic properties of the soundtrack playing in the background, *Food Qual. Pref.* **24**, 201–204.

Davis, N. (2015). Welcome to the Tate Sensorium, where the paintings come with chocolates, *The Guardian*, **August 22nd**. http://www.theguardian.com/artanddesign/video/2015/aug/25/welcome-tate-sensorium-taste-touch-smell-art-video.

Demattè, M. L., Pojer, N., Endrizzi, I., Corollaro, M. L., Betta, E., Aprea, E., Charles, M., Biasioli, F., Zampini, M. and Gasperi, F. (2014). Effects of the sound of the bite on apple perceived crispness and hardness, *Food Qual. Pref.* **38**, 58–64.

Feinstein, A. H., Hinskton, T. S. and Erdem, M. (2002). Exploring the effects of music atmospherics on menu item selection, *J. Foodservice Bus. Res.* **5**(4), 3–25.

Fritz, T. H., Brummerloh, B., Urquijo, M., Wegner, K., Reimer, E., Gutekunst, S., Schneider, L., Smallwood, J. and Villringer, A. (2017). Blame it on the bossa nova: transfer of perceived sexiness from music to touch, *J. Exp. Psychol. Gen.* **146**, 1360–1365.

Gourévitch, B., Edeline, J.-M., Occelli, F. and Eggermont, J. J. (2014). Is the din really harmless? Long-term effects of non-traumatic noise on the adult auditory system, *Nature Rev. Neurosci.* **15**, 483–491.

Guéguen, N., Le Guellec, H. and Jacob, C. (2004). Sound level of background music and consumer behavior: an empirical evaluation, *Percept. Mot. Skills* **99**, 34–38.

Guéguen, N., Jacob, C., Le Guellec, H., Morineau, T. and Lourel, M. (2008). Sound level of environmental music and drinking behavior: a field experiment with beer drinkers, *Alcohol Clin. Exp. Res.* **32**, 1795–1798.

Hauck, P. and Hecht, H. (2019). Having a drink with Tchaikovsky: the crossmodal influence of background music on the taste of beverages, *Multisens. Res.* **32**, 1–24.

Höchenberger, R. and Ohla, K. (2019). A bittersweet symphony: evidence for taste-sound correspondences without effects on taste quality-specific perception, *J. Neurosci. Res.* **97**, 267–275.

Kaiser, D., Silberger, S., Hilzendegen, C. and Stroebele-Benschop, N. (2016). The influence of music type and transmission mode on food intake and meal duration: an experimental study, *Psychol. Music* **44**, 1419–1430.

Kantono, K., Hamid, N., Sheperd, D., Yoo, M. J. Y., Carr, B. T. and Grazioli, G. (2015). The effect of background music on food pleasantness ratings, *Psychol. Music* **44**, 1111–1125. DOI:10.1177/0305735615613149.

Kantono, K., Hamid, N., Sheperd, D., Yoo, M. J. Y., Grazioli, G. and Carr, T. (2016). Listening to music can influence hedonic and sensory perceptions of gelati, *Appetite* **100**, 244–255.

Kantono, K., Hamid, N., Shepherd, D., Lin, Y. H. T., Skiredj, S. and Carr, B. T. (2019). Emotional and electrophysiological measures correlate to flavour perception in the presence of music, *Physiol. Behav.* **199**, 154–164.

Lin, Y. H. T., Hamid, N., Shepherd, D., Kantono, K. and Spence, C. (2019). Environmental sounds influence the multisensory perception of chocolate gelati, *Foods* **8**(4), 124.

Lowe, M., Ringler, C. and Haws, K. (2018). An overture to overeating: the cross-modal effects of acoustic pitch on food preferences and serving behaviour, *Appetite* **123**, 128–134.

Muniz, R., Harrington, R. J., Ogbeidea, G.-C. and Seo, H.-S. (2017). The role of sound congruency on ethnic menu item selection and price expectations, *Int. J. Hosp. Tour. Admin.* **18**, 245–271.

Novak, C. C., La Lopa, J. and Novak, R. E. (2010). Effects of sound pressure levels and sensitivity to noise on mood and behavioral intent in a controlled fine dining restaurant environment, *J. Culin. Sci. Technol.* **8**, 191–218.

Pettit, L. A. (1958). The influence of test location and accompanying sound in flavor preference testing of tomato juice, *Food Technol.* **12**, 55–57.

Pursey, T. and Lomas, D. (2018). Tate Sensorium: an experiment in multisensory immersive design, *The Senses and Society* **13**(3), 354–366.

Rahne, T., Köppke, R., Nehring, M., Plontke, S. K. and Fischer, H. G. (2018). Does ambient noise or hypobaric atmosphere influence olfactory and gustatory function?, *PLoS One* **13**(1), e0190837.

Reinoso Carvalho, F., Van Ee, R., Rychtarikova, M., Touhafi, A., Steenhaut, K., Persoone, D., Spence, C. and Leman, M. (2015). Does music influence the multisensory tasting experience?, *J. Sens. Stud.* **30**, 404–412.

Richard, G. (2012). Eating out may be bad for your ears, *The Washington Post*, **July 7th**.

Royce-Greensill, S. (2017). The Berkeley launches Out Of The Blue: review, *The Telegraph*, **November 17th**. https://www.telegraph.co.uk/luxury/drinking-and-dining/berkeley-blue-cocktail-tasting-review/.

Septianto, F. (2016). "Chopin" effect? An exploratory study on how musical tempo influence consumer choice of drink with different temperatures, *Asia Pac. J. Mark. Logist.* **28**, 765–779.

Spence, C. (2011). Crossmodal correspondences: a tutorial review, *Atten. Percept. Psychophys.* **73**, 971–995.

Spence, C. (2012). Auditory contributions to flavour perception and feeding behaviour, *Physiol. Behav.* **107**, 505–515.

Spence, C. (2014a). Noise and its impact on the perception of food and drink, *Flavour* **3**, 9. DOI:10.1186/2044-7248-3-9.

Spence, C. (2014b). Assessing the influence of shape and sound symbolism on the consumer's response to chocolate, *N. Food* **17**, 59–62.

Spence, C. (2015). Eating with our ears: assessing the importance of the sounds of consumption to our perception and enjoyment of multisensory flavour experiences, *Flavour* **4**, 3. DOI:10. 1186/2044-7248-4-3.

Spence, C. (2017a). Sonic seasoning, in: *Audio Branding: Using Sound to Build Your Brand*, L. Minsky and C. Fahey (Eds), pp. 52–58. Kogan Page, London, UK.

Spence, C. (2017b). *Gastrophysics: the New Science of Eating*. Viking Penguin, London, UK.

Spence, C. (2019a). Multisensory experiential wine marketing, *Food Qual. Pref.* **71**, 106–116.

Spence, C. (2019b). On the relative nature of (pitch-based) crossmodal correspondences, *Multisens. Res.* **32**, 235–265.

Spence, C. and Gallace, A. (2011). Multisensory design: reaching out to touch the consumer, *Psychol. Mark.* **28**, 267–308.

Spence, C. and Wang, Q. (J.) (2015). Sensory expectations elicited by the sounds of opening the packaging and pouring a beverage, *Flavour* **4**, 35. DOI:10.1186/s13411-015-0044-y.

Spence, C., Richards, L., Kjellin, E., Huhnt, A.-M., Daskal, V., Scheybeler, A., Velasco, C. and Deroy, O. (2013). Looking for crossmodal correspondences between classical music & fine wine, *Flavour* **2**, 29. DOI:10.1186/2044-7248-2-29.

Spence, C., Shankar, M. U. and Blumenthal, H. (2011). 'Sound bites': auditory contributions to the perception and consumption of food and drink, in: *Art and the Senses*, F. Bacci and D. Melcher (Eds), pp. 207–238. Oxford University Press, Oxford, UK.

Velasco, C., Reinoso Carvalho, F., Petit, O. and Nijholt, A. (2016). A multisensory approach for the design of food and drink enhancing sonic systems, in: *Proceedings of the 1st Workshop on Multi-Sensorial Approaches to Human–Food Interaction (MHFI'16)*, A. Nijholt, C. Velasco, G. Huisman and K. Karunanayaka (Eds), Article 7. ACM, New York, NY, USA.

Victor, A. (2014). Louis Armstrong for starters, Debussy with roast chicken and James Blunt for dessert: British Airways pairs music to meals to make in-flight food taste better, *Daily Mail Online*, **October 15th**. http://www.dailymail.co.uk/travel/travel_news/article-2792286/british-airways-pairs-music-meals-make-flight-food-taste-better.html.

Wang, Q. (J.) and Spence, C. (2016). "Striking a sour note": assessing the influence of consonant and dissonant music on taste perception, *Multisens. Res.* **30**, 195–208.

Wang, Q. (J.) and Spence, C. (2018). "A sweet smile": the modulatory role of emotion in how extrinsic factors influence taste evaluation, *Cogn. Emot.* **32**, 1052–1061.

Wang, Q. J. and Spence, C. (2019). Sonic packaging: how packaging sounds influence multisensory product evaluation, in: *Multisensory Packaging: Designing New Product Experiences*, C. Velasco and C. Spence (Eds), pp. 103–125. Palgrave MacMillan, Cham, Switzerland.

Watson, Q. J. and Gunter, K. L. (2017). Trombones elicit bitter more strongly than do clarinets: a partial replication of three studies of Crisinel and Spence, *Multisens. Res.* **30**, 321–335.

Woods, A. T., Poliakoff, E., Lloyd, D. M., Kuenzel, J., Hodson, R., Gonda, H., Batchelor, J., Dijksterhuis, G. B. and Thomas, A. (2011). Effect of background noise on food perception, *Food Qual. Pref.* **22**, 42–47.

Yan, K. S. and Dando, R. (2015). A crossmodal role for audition in taste perception, *J. Exp. Psychol. Hum. Percept. Perform.* **41**, 590–596.

Youssef, J., Youssef, L., Juravle, G. and Spence, C. (2017). Plateware and slurping influence regular consumers' sensory discriminative and hedonic responses to a hot soup, *Int. J. Gastron. Food Sci.* **9**, 100–104.

Zampini, M. and Spence, C. (2004). The role of auditory cues in modulating the perceived crispness and staleness of potato chips, *J. Sens. Sci.* **19**, 347–363.

Zellner, D., Geller, T., Lyons, S., Pyper, A. and Riaz, K. (2017). Ethnic congruence of music and food affects food selection but not liking, *Food Qual. Pref.* **56**, 126–129.

Extrinsic Auditory Contributions to Food Perception & Consumer Behaviour: an Interdisciplinary Review

Charles Spence [1,*], **Felipe Reinoso-Carvalho** [2], **Carlos Velasco** [3] and **Qian Janice Wang** [4]

[1] Department of Experimental Psychology, Anna Watts Building, University of Oxford, Oxford, OX2 6GG, UK
[2] School of Management, Los Andes University, Bogotá, Colombia
[3] BI Norwegian Business School, Norway
[4] Department of Food Science, Aarhus University, Denmark

Abstract

Food product-extrinsic sounds (i.e., those auditory stimuli that are not linked directly to a food or beverage product, or its packaging) have been shown to exert a significant influence over various aspects of food perception and consumer behaviour, often operating outside of conscious awareness. In this review, we summarise the latest evidence concerning the various ways in which what we hear can influence what we taste. According to one line of empirical research, background noise interferes with tasting, due to attentional distraction. A separate body of marketing-relevant research demonstrates that music can be used to bias consumers' food perception, judgments, and purchasing/consumption behaviour in various ways. Some of these effects appear to be driven by the arousal elicited by loud music as well as the entrainment of people's behaviour to the musical beat. However, semantic priming effects linked to the type and style of music are also relevant. Another route by which music influences food perception comes from the observation that our liking/preference for the music that we happen to be listening to carries over to influence our hedonic judgments of what we are tasting. A final route by which hearing influences tasting relates to the emerging field of 'sonic seasoning'. A developing body of research now demonstrates that people often rate tasting experiences differently when listening to soundtracks that have been designed to be (or are chosen because they are) congruent with specific flavour experiences (e.g., when compared to when listening to other soundtracks, or else when tasting in silence). Taken together, such results lead to the growing realization that the crossmodal influences of music and noise on food perception and consumer behaviour may have some important if, as yet, unrecognized implications for public health.

---

[*] To whom correspondence should be addressed. E-mail: charles.spence@psy.ox.ac.uk

© KONINKLIJKE BRILL NV, LEIDEN, 2019 | DOI:10.1163/9789004416307_004

Keywords
Auditory, chemical senses, food, noise, crossmodal, multisensory, taste, flavour

1. Introduction

What we hear affects what we taste, no matter whether we realise it or not
(and the evidence suggests that mostly we do not, e.g., see North *et al.*, 1997,
1999; Zellner *et al.*, 2017). In fact, there is now an extensive body of literature
highlighting the impact of the sounds that may be associated with food preparation (Wheeler, 1938; see Knöferle and Spence, in press, for a recent review),
food packaging (i.e., being opened; Spence and Wang, 2015a, 2017; see Wang
and Spence, 2019, for a review), and food consumption (e.g., Youssef *et al.*,
2017; Zampini and Spence, 2004; see Spence, 2015a, for a review), on people's sensory-discriminative and hedonic ratings of a wide range of different
food and drink products. Such product-intrinsic auditory contributions to food
perception and consumer behaviour are undoubtedly important. However, the
focus of the present review will be squarely on the effect of product-extrinsic
sounds on what we taste, broadly construed.

In what is perhaps the earliest work in this area, Petitt (1958) had her participants taste and rate tomato juice, though no effect of modest levels of background noise was observed. However, despite such an inauspicious start some
70 years ago, research on the auditory contributions to food perception and
consumer behaviour has exploded in recent years, thus necessitating an up-to-
date review of the literature, as provided here. The topic has sparked interest
in a diverse range of fields that include experimental psychology, cognitive
neuroscience, design, music, marketing, gastronomy, branding, and beyond.
Indeed, an extensive body of research published over the last half century or
so has now convincingly demonstrated that the background sounds and music
that happen to be playing in bars, restaurants, cafes, and stores bias what customers choose to purchase, order, and/or consume, not to mention what they
think it tastes like, how much they enjoy — and would be willing to pay for —
the experience (e.g., Biswas *et al.*, 2019; Reinoso Carvalho *et al.*, 2019; see
Spence, 2017, for a review).

In the following sections, we review the evidence concerning four of the
main ways in which what we hear, despite being seemingly unrelated to what
we are tasting, can nevertheless still influence our perception of food and
drink, as well as modifying various food-related consumer behaviours. We
start by assessing the very general, and relatively stimulus non-specific, effects of background noise on tasting. Next, we assess the effects of background
music on food perception and consumer behaviour. We review the effects of
loud music on arousal, as well as briefly summarize the evidence showing that

consumers' (food and beverage-related) behaviour is often entrained to the musical beat. In this section, we also look at those priming effects that appear to be associated with the type of music, as well as any other associations that may be primed musically in the mind of the consumer. Thereafter, we take a look at the phenomenon of 'sensation transference', sometimes referred to as 'affective ventriloquism' or the 'halo effect'. This is where our liking for whatever we are listening to carries over to influence our judgment of whatever we happen to be tasting. Finally, we review the rapidly evolving literature documenting the much more stimulus-specific effects of 'sonic seasoning' on multisensory tasting experiences.

While there have been a number of previous reviews summarizing various aspects of audition's interaction with/influence over tasting, and even a couple covering the same broad areas outlined here, it seems timely for an update given the sheer number of recently-published papers on the topic of sonic seasoning. This review also includes a recently unveiled model summarizing the way in which sonic seasoning might work, as well as providing a new analysis of experiment designs and effect sizes in this area of research.

Taken together, such crossmodal effects can be seen as particularly intriguing, given that the auditory stimuli concerned have no direct connection with food or drink (see Spence and Deroy, 2013a). In all such cases, the noise, music, or the especially composed soundscape, are extrinsic to the food products under consideration. This certainly contrasts with, e.g., the sound of a sizzling steak as it arrives at the table (Wheeler, 1938), the crunch of a celery stick in the mouth, or the pop of the champagne cork as it leaves the bottle (see Spence, 2015a, for a review). At the outset, though, it is perhaps worth highlighting the fact that, while the four above-mentioned broad areas of research have remained relatively segregated in the academic literature over the decades, there are grounds for thinking that the distinctions between them may not always be as clear-cut as it at first may seem, especially at the boundaries. So, for example, think here only of how background music turns into 'noise' if played at a 'too loud' level. Similarly, one might also wonder whether the matching of types (or ethnicities) of music with types (or ethnicities) of cuisine (see Reinoso Carvalho *et al.*, 2016a, for evidence on this score) is not itself an example of a high-level crossmodal correspondence, one that is in some ways akin to the sonic seasoning we cover in a later section (see section 5). We will address these uncertainties as they arise in the sections below.

2. Background Noise and Its Impact on Tasting

When what we hear becomes too loud, we usually frame it as 'noise', and the possibly detrimental effect of noise is perhaps the oldest concern of researchers working on the influence of sound on tasting (see Crocker, 1950;

Petitt, 1958; Srinivasan, 1955, for early discussion and research). It is also perhaps the most nonspecific of product-extrinsic auditory stimuli in terms of its impact on food perception. While complaints about noise in restaurants and bars would appear to have been on the rise in the west in recent years (e.g., Belluz, 2018; Moir, 2015; see Spence, 2014a, for a review), it is worth noting that researchers have actually been commenting on overly loud restaurants for many decades now (see Petitt, 1958, for an early example). The research that has been published to date shows that loud background noise, regardless of whether it is airplane noise, white noise, or even the background noise of a restaurant, or bar, affects both the perceived taste of food and drink, as well as people's ability to discriminate various aspects of their tasting experience (Rahne *et al.*, 2018; Trautmann *et al.*, 2017; see Spence, 2014a, for a review).

At around the same time as Petitt (1958) published her early research, other commentators were suggesting that loud background noise distracted from tasting and/or interfered with the tasting experience (see Crocker, 1950; Peynaud, 1987) (see Note 1). Crucially, a series of empirical studies conducted over the last decade has illustrated the interfering effect of loud background noise on both tasting and smelling (see also Novak, La Lopa and Novak, 2010). For example, using a range of everyday foods, Woods *et al.* (2011a) demonstrated that the ability of untrained participants (tested in the UK) to taste sweet and salt, as well as their perception of crunchy food, was suppressed under the influence of loud background white noise (in this case, presented over headphones at around 80–85 dB). The foods tasted in this study consisted of typical snack foods, such as Pringles Original Salted Crisps and Sainsbury's Nice Biscuits. Meanwhile, Yan and Dando (2015; building on predictions made by Spence *et al.*, 2014a), reported that ratings of the subjective intensity of the five basic tastants (sweet, salty, sour, bitter, and umami) presented in solution were, in several cases, affected when accompanied by airplane noise at 80–85 dB (i.e., set at roughly the same level one would be exposed to in a commercial airplane). In particular, ratings of sweetness were suppressed significantly, while the umami solution was rated as tasting more intense amongst their North American participants (Note 2). Interestingly, this may help to explain why so many passengers seem to choose a tomato juice, or a Bloody Mary, while on an airplane (see Spence, 2017, for a review) (Note 3).

Research by Seo *et al.* (2012) has also shown that background noise can, at least under certain conditions, influence people's sensitivity to odours (see also Seo *et al.*, 2011). So, for example, Seo *et al.* (2011) played various kinds of background noise over headphones to participants who were performing an odour discrimination task. The participants had to pick the odd one out of three "Sniffin' sticks" (odorous felt-tip pens), two of which had the same odour, while the remaining one smelled differently. Verbal noise, consisting of someone reading an audio book at 70 dB, exerted more of a detrimental effect on participants' performance than party noise presented at the same level, which, in turn, was more detrimental than silence. By contrast, listening to Mozart's sonata for two pianos in D major K448 did not affect performance relative to a silent baseline condition.

In a follow-up study, Seo *et al.* (2012) showed that performance on an odour sensitivity task wasn't affected by the presence of background noise (either verbal or non-verbal) when compared to a baseline silent condition. However, that said, in this case, a closer look at the data revealed that while verbal background noise significantly impaired the olfactory sensitivity of introverted participants, it had the opposite effect on the more extroverted participants. Elsewhere, Velasco *et al.* (2014) instructed participants to rate six food-related odours (lemon, orange, bilberry, musk, dark chocolate, and smoked) while either listening to music or white noise (once again presented over headphones at 70 dB). These olfactory stimuli were rated as significantly less pleasant (by around 5%) in the presence of white noise than when either pleasant or unpleasant (consonant and dissonant) musical selections were played instead.

By-and-large, the results reported in this section would therefore appear consistent with the suggestion that loud background noise acts as a crossmodal distractor or masking stimulus (e.g., see Hockey, 1970; Kou, McClelland and Furnham, 2018; Plailly *et al.*, 2008; Spence, 2014a; see also Wesson and Wilson, 2010, 2011) (Note 4). What is also still unclear is why noise suppresses our perception of certain attributes of the tasting experience while at the same time seemingly boosting others (e.g., umami). According to one evolutionary argument (Ferber and Cabanac, 1987), building on early work in the animal model (Kupferman, 1964), the suggestion has been forwarded that in times of stress, such as when exposed to loud noise, we may find those tastes that signal energy (e.g., sweetness) to be more palatable, the idea here being that such changes might serve an evolutionarily useful function in helping an organism to secure sufficient energy in order to deal with the stressful situation. However, even though such a suggestion may sound intriguing, convincing evidence in support of this notion has yet to be forthcoming.

3. Background Music

In this section, we move on from looking at the effects of background noise (be it defined as nonspecific, or unpleasant, type of sound), to a consideration of the impact that background music has both on consumer behaviour and food perception. The section is broken into three broad classes of crossmodal influence. We start with the effect of loud music on consumption, possibly mediated by arousal. Next, we take a brief look at the behavioural entrainment to the musical beat that has been reported in various food-related consumption contexts. Finally, we examine the semantic priming effects that are elicited as a function of the type of music that the consumer is exposed to.

3.1. Loud Music

The laboratory research that has been published to date demonstrates that increasing the loudness of the background music results in participants drinking more (e.g., McCarron and Tierney, 1989). Crucially, real-world studies have also confirmed that consumers tend to drink more when the volume of the background music is turned up (Guéguen *et al.*, 2004, 2008). In fact, according to a report that appeared in *The New York Times*, the Hard Rock Café chain deliberately plays loud music because of the positive effect it has on sales (Note 5). Just take the following quote from the newspaper article itself: *"[T]he Hard Rock Café had the practice down to a science, ever since its founders realized that by playing loud, fast music, patrons talked less, consumed more and left quickly, a technique documented in the International Directory of Company Histories."* (Buckley, 2012). Meanwhile, according to Clynes (2012): *"When music in a bar gets 22 per cent louder, patrons drink 26 per cent faster."* Music that is very loud is sometimes also used in order to deter a certain profile of customers from drinking/dining in a particular venue (Forsyth and Cloonan, 2008).

Nowadays, there would appear to be a growing groundswell of opinion suggesting that many restaurants/bars in North America, the UK, Australia, and beyond, are becoming louder (see Spence, 2014a, for a review of this literature). This is not solely due to chefs/restaurateurs speculating that loud music in the dining room is somehow a good idea (see Spence, 2015b). Rather, part of the 'blame' here should fall at the doors of those who prioritize the modern design aesthetic, whereby many of the sound-absorbing soft furnishings (curtains, cushions, and carpets) are replaced with 'minimalist' hard reflective surfaces (see Spence and Piqueras-Fiszman, 2014).

Stafford and his colleagues (Stafford *et al.*, 2012, 2013) have demonstrated that people find it harder to discern the alcohol content of drinks under conditions of loud background noise (Note 6). In particular, in 2012, Stafford *et al.* reported that their participants ($N = 80$) rated alcoholic beverages as tasting sweeter when listening to loud background music (comprising Drum and Bass, House, Hardcore, Dubstep, and Trance) than in the absence of background music. These results, note, seemingly contradict those obtained by Woods *et al.* (2010), reported earlier, in the sense that opposite effects on sweetness perception were documented in the two studies as a result of participants being subjected to loud sound.

Ultimately, of course, the most appropriate music loudness level may depend on the style of a given venue. So, for instance, 80 diners in one North American study spent around 15% more when quieter, as opposed to louder, background classical, or soft rock music, was playing (Lammers, 2003). In this case, it was suggested that the quieter the music, the better match with the 'serene' atmosphere of this ocean-side California restaurant.

The fact that listening to loud background music so often increases consumption may be attributable to the impact that music has on arousal. Music can, after all, be used to arouse or relax people (e.g., North and Hargreaves, 1997), with the suggestion here that people tend to consume more when they are more aroused. There may, of course, be social and societal factors relevant to the consumption of certain drinks (e.g., alcohol) in terms of social desirability, for instance, when in the presence of music. Alternatively, however, the effect of loud music might also reflect some kind of state-dependent learning/behaviour. Assuming that what people normally do at parties where the music is loud is drink, and eat, reinstating such sensory environmental cues may simply help to prime the associated behaviour (cf. Remington *et al.*, 1997). There is also likely a conditioning angle to the impact of auditory stimuli on the consumer (see also Hernando, 2014). After all, Pavlov's dogs learned to associate a food-unrelated auditory cue (the ding of the bell) with the appearance of food, and hence started to salivate in response to the sound as a result (Pavlov, 1921/1927). Intriguingly, similar associative learning effects have also been demonstrated in fish (Frolov, 1937) (Note 7).

Given the increasing noise levels in many restaurants and bars these days, there would seem to be a possible public health angle to this research as well (Note 8). As a case in point, Biswas *et al.* (2019) have recently published research showing that low volume background music/noise leads to an increased sale of healthy foods compared to high volume or no music/noise, the suggestion being that this was presumably due to the sense of relaxation that was induced in the shoppers. In contrast, high volume music/noise results in increased levels of excitement (what one might think of as increased arousal), and this leads to an increase in the purchase of unhealthy foods. The role of music in nudging healthful behaviour is something we would like to highlight in this review, and we will return to later.

3.2. Musical Tempo

Several studies have demonstrated that a range of consumer behaviours tend to become somewhat entrained toward the tempo of the background music (Roballey *et al.*, 1985; see also Knoeferle *et al.*, 2017). For instance, participants in laboratory studies drink more rapidly when high- (rather than low-) tempo music is played. Similar results have also been documented in more ecologically-valid studies conducted in a variety of bars and restaurants (e.g., Bach and Schaefer, 1979; Caldwell and Hibbert, 2002; Milliman, 1986). For instance, in one of the largest studies of its kind, Milliman reported a 30% increase in average dollar spend on the bar tab amongst 1400 diners when slow, rather than fast, tempo music was played. Milliman hypothesised that the slower tempo music may have encouraged the diners to linger for longer. That some food chains really do try to control the flow of customers through their premises, is suggested by the following quote from Chris Golub, the man responsible for selecting the music that plays in all 1500 Chipotle branches in the US: *"The lunch and dinner rush have songs with higher BPM because they need to keep the customers moving."* (quoted in Suddath, 2013). It is worth thinking about the public health implications here: To the extent that people chew faster and/or for less time before swallowing in the presence of loud music, this is likely to have an impact on satiety, possible also subsequently on digestion, and hence eventually on consumption. That said, we are not aware of any carefully-controlled empirical evidence on this score.

3.3. Musical Style

The type, or style, of music that happens to be playing in the background has been shown to exert a surprisingly pronounced effect on consumer choice behaviour in a range of real-world environments (e.g., see North *et al.*, 1997, 1999; Zellner *et al.*, 2017). The type or style of music has also been shown to influence what people have to say about the tasting experience itself (e.g., North, 2012; Yeoh and North, 2010). Here, though, one might want to distinguish between those associations that may be primed by the sonic attributes of the music, and the more complex sematic associations that may be primed by the style of music (be it, for instance, ethnic or classical music — Hutchison, 2003; Labroo *et al.*, 2008; Lucas, 2000; see also Muniz *et al.*, 2017).

In their now classic study, North *et al.* (1997, 1999) demonstrated a marked reversal in sales of French and German wine in a British supermarket as a function of whether French accordion *vs* German Bierkeller music happened to be playing in the background. What is more, only six of the 44 consumers who agreed to be questioned after leaving the tills thought that the atmospheric music had influenced their purchasing behaviour. More recently, Zellner *et al.* (2017) demonstrated that people ($N = 275$ North American students and faculty) given a choice of Spanish *vs* Italian meals (seafood paella *vs* chicken parmesan; or other dishes) in a university canteen were significantly more likely to choose the paella when instrumental Spanish, rather than Italian, music was playing (34% *vs* 17%, respectively). Once again, the majority of diners (82 out of the 84 interviewed afterwards) denied that the background music had influenced their meal choice. No effect of musical congruency on hedonic responses to the chosen dish was reported in this study (cf. Yeoh and North, 2010, for weak evidence). However, it is worth noting that this latter null result may simply reflect the fact that (as Zellner and her colleagues themselves readily acknowledged) the background music was not especially (or even necessarily) audible in the dining area where the hedonic ratings were made in this study. Other laboratory research, meanwhile, has demonstrated that the type (or genre) of background music can modulate flavour pleasantness and people's overall impression of various food stimuli (Fiegel *et al.*, 2014; see also Martens *et al.*, 2010). One possibility here, of course, is that the style of music might bias the eye movements and visual search behaviour of consumers (cf. Knoeferle *et al.*, 2016, for evidence concerning visual search biased by sonic logos).

A number of real-world studies have shown that playing background classical music (e.g., when compared to Top-40 hits) leads to consumers spending more on their food and beverage purchases, no matter whether they happen to be in a wine shop (Areni and Kim, 1993), a university cafeteria (North and Hargreaves, 1998; North *et al.*, 2003, 2016), or even an African-themed restaurant (Wilson, 2003). The suggestion that is often put forward here is that playing classical music semantically primes notions of quality and class, which nudges consumers into spending more than they otherwise might (see also Magnini and Thelen, 2008). At the same time, however, it is perhaps also worth pointing out how classical music can be used as a deterrent. For instance, McDonalds plays classical music outside a number of their more popular 24-h inner city establishments in order to try and reduce the likelihood of youths gathering (Taylor, 2017), classical music being semantically incongruent with most people's notion of what McDonalds stands for. Note that something very similar was done several decades earlier outside one store chain's premises in California. According in Lanza (2004, pp. 226–227): "The Los Angeles Times reported that the Southland Corporation had installed a Muzak channel in its 7–11 chain stores in Southern California's Thousand Oaks district. The goal was to drive away gangs of loitering teenagers. The plan worked so well that the company wanted to repeat the 'Muzak Attack' in other parts of LA county. It was actually Muzak's classical channel."

North (2012) conducted a study showing that background music can be used to prime, and hence bias, attributes of the tasting experience, such as assessments of how 'powerful and heavy' or 'zingy and refreshing' a wine appears to be. In his study, North had 250 students studying in Scotland evaluate a glass of either white or red wine, while at the same time listening to music that had been pre-determined to be associated with one of four metaphorical categories ('powerful and heavy', 'zingy and refreshing', 'subtle and refined', and 'mellow and soft'). The students' judgments of the wine were influenced by the music, with the students rating both wines as tasting more 'powerful and heavy' when listening to *Carmina Burana* by Karl Orff, and as tasting more zingy and refreshing when listening to *Nouvelle Vague's* "Just Can't get Enough". While it is assimilation effects such as these that are normally reported, there is an open question here as to whether contrast effects might also be documented as well under the appropriate conditions (see Piqueras-Fiszman and Spence, 2015, for a review).

4. Sensation Transference

Over the years, a number of researcher have addressed the question of whether *"If you like the music more, do you like what you are eating/drinking more too?"* (e.g., Kantono *et al.*, 2016a, b, 2018). Such crossmodal effects can be thought of as an example of 'sensation transference'. Seo and Hummel (2011) have also reported transfer effects, showing that auditory cues can modulate odour pleasantness (see also Seo and Hummel, 2011, 2015; Seo *et al.*, 2014). In their 2009 study, for example, Seo and Hummel demonstrated that the hedonic valence associated with auditory stimuli can transfer to the odours, and that such transference doesn't seem to be dependent on people hedonically evaluating the odour.

It is, though, currently an open question as to whether sensation transference effects may also be observed for other attributes such as, for example, arousal (see Spence and Wang, 2015c). Indeed, elsewhere in the literature, it is clear that sensation transference effects do not necessarily occur between all pairs of stimuli/stimulus dimensions (e.g., see Fritz *et al.*, 2017; Marin *et al.*, 2017, for a couple of examples).

Reinoso-Carvalho *et al.* (2019) conducted a series of recent experiments in which consumers tasted and rated one of a range of beers while listening to either a positively (or negatively) valenced piece of music. In these experiments, participants generally liked the beer more, and rated it as tasting sweeter, when listening to music having positive, as compared to negative, emotion (Note 9). The same beer was rated as tasting more bitter, as having a higher alcohol content, and as having more body when experienced with the music having negative, as compared to positive, emotion. Importantly, from a marketing perspective, the participants in this study were also willing to pay 7–8% more for the same beer tasted while listening to positive, as compared to negative, music. Meanwhile, in another recent study, Ziv (2018) reported that cookies were rated as tasting better when people listened to pleasant background music. Interestingly, however, in this study a larger difference in the evaluation of the cookies was observed when the first cookie was tasted with pleasant (as compared to unpleasant) background music (see also Kantono *et al.*, 2016c). In another example linking physiological measures, self-rated emotion, and perceived tastes, participants listened to liked, disliked, and neutral music while rating gelato using the method of temporal dominance of sensations (Kantono *et al.*, 2019). The authors found that positive emotions were associated with the dominance of sweet and milky tastes/flavours whereas negative emotions were associated with bitter and creamy tastes/flavours instead.

It might be suggested that the sensation transference effects that have been reported so far in this section can be considered as a kind of 'affective priming'. According to such a view, the only difference from the results reported in the previous section is that what is being primed is valence rather than the type (i.e., ethnicity or class) of music (Note 10). Note here that when sensation transference relates specifically to valence, it is also described as the halo effect (Clark and Lawless, 1994) and affective ventriloquism (see Spence and Gallace, 2011). Here, though, there is uncertainty as to whether it is what people think about the music that is being transferred to what they think about what they are tasting. Alternatively, however, one might also argue that the emotion conveyed by the music influences the emotional state of the taster, and it is that, that affects their taste ratings (see Konečni, 2008). Elsewhere, after all, it has been shown that sweetness is rated as more intense (while sourness is rated as less intense) by those tasting after their hockey team has won, as compared to the ratings given when the fan's team has just lost (Noel and Dando, 2015). Such results would appear to provide some support for the latter account. However, presumably, these explanations should not be considered as being mutually exclusive. It is also important to note here that sensation transference is certainly not restricted just to music. In a crossmodal study involving both visual and auditory stimuli with matched valence, Wang and Spence (2018) recently demonstrated that participants rated juice samples as tasting sweeter and less sour when they were exposed to pleasant stimuli, regardless of whether they saw images of a happy (*vs* sad) face or listened to consonant (*vs* dissonant) music.

Congruent music may, of course, affect people's responses to the service environment too (i.e., and not just the food and/or drink served in a particular environment). In turn, what the diner thinks about the environment may then itself result in sensation transference which biases people's ratings of the food/drink. So, for instance, Demoulin (2011) investigated the impact of congruent musical choices on the emotional and cognitive responses of diners to the environment (specifically a healthy fast-food restaurant in France offering balanced meals with quality products and trendy recipes). Musical congruency, as assessed by a small number of the restaurant's regular customers (congruent music was described as 'modern, pop and dynamic' whereas the incongruent music was made up of 'old-fashioned timeless hits') led to lower arousal and increased pleasure. This, in turn, increased customers' evaluation of the environment quality and service quality. This, then, provides another example of the way in which the environment 'as a whole' may have an impact on food evaluation, though the lines between sensation transference and crossmodal congruency/correspondences are sometimes blurred.

One other question to consider here is what exactly the difference is between hedonic 'sensation transference' and those crossmodal correspondences that would appear to be mediated by affect (see section 5). It is not clear that anyone has a good answer here yet, but it is perhaps nevertheless still worth bearing this in mind as one of the blurry boundaries between the four ways in which sound affects food perception that have been outlined here.

5. Crossmodal Correspondences Between Audition and the Chemical Senses

A recently discovered fourth route by which what we hear can influence what we taste is based on the notion of 'sonic seasoning'. This is where pieces of music, or soundscapes, are especially chosen, or even composed, in order to correspond crossmodally with the taste, aroma, mouthfeel, or flavour of a particular food or drink (see Table 1 for an overview of recent studies demonstrating sonic seasoning).

To be clear, crossmodal correspondences are defined as the connections that many people appear to experience between features, attributes, and/or dimensions of experience in different sensory modalities that do not share anything obviously in common (see Parise and Spence, 2013; Spence, 2011). It is because they initially seem so surprising that people often consider them, incorrectly in our opinion, as a kind of synaesthesia (see Deroy and Spence, 2013). Interesting questions here concern where such surprising correspondences come from (Note 11), and the conditions under which corresponding/congruent *vs* incongruent (or no) music influences the tasting experience (e.g., Hauck and Hecht, 2019; Höchenberger and Ohla, 2019; Spence and Deroy, 2013a; Watson and Gunter, 2017).

The earliest studies in this area by Kristan Holt-Hansen (1968, 1976) provided some initial evidence that people ($N = 16$) associated a higher-pitched pure tone (640–670 Hz *vs* 510–520 Hz) with a beer that was more alcoholic, and that drinking the beer while listening to the matching tone led to higher pleasantness ratings for at least some of the participants. A few years later, Rudmin and Capelli (1983) partially replicated these results and extended them to a broader range of foods including the same beers, plus non-alcoholic beer, grapefruit juice, hard candy, and dill pickle. The small sample of participants ($N = 10$) chose significantly higher frequencies for the acidic foods (grapefruit juice, candy, pickle) compared to the beers. More recently, still, we have extended this approach to matching with a range of Belgian beers and other drinks (e.g., Reinoso Carvalho *et al.*, 2016a, b, c), not to mention with sample sizes that are much larger.

For a more systematic approach, one should perhaps consider simpler gustatory stimuli consisting of basic tastes. A series of tests involving basic tastes was conducted by Anne-Sylvie Crisinel at the Crossmodal Research Laboratory at Oxford. Implicit Association Tests revealed an association between

Table 1.

A summary of recent studies demonstrating sonic seasoning *via* the use of soundtracks/music (rather than product-induced sounds). Effect size (Cohen's *d*) provided where data are available for calculations. Cohen's *d* provides a measure of effect size indicating standardised difference between two means, which allows for comparison of effect sizes across different studies. Percent difference refers to the differences in attributes between the sound conditions listed under auditory stimuli. In the case of more than two soundtracks, explicit comparison conditions are listed in parentheses

Study	Auditory stimuli	Food/drink	DV	Study design	Sample size	% difference	Effect size (Cohen's *d*)
Crisinel *et al.*, 2012	Sweet, bitter soundtracks	Cinder toffee	9-point scales: sweet–bitter, position, liking	Within participants	20	15% sweeter	0.5
North, 2012	4 pieces of music + silence	Wine (1 white and 1 red)	11-point scales: powerful/heavy, subtle/refined, zingy/refreshing, mellow/soft, wine liking	Between participants	250 (25 per cell)	40% more zingy/fresh, 32% more powerful/heavy, 29% more mellow/soft, 30% more subtle/refined (each soundtrack compared against all other conditions)	
Spence *et al.*, 2013, study 2	Classical music matching wines, silence	Wine (1 white, 2 red)	11-point scales: sweetness, acidity, alcohol, fruit, tannin, enjoyment	Within participants	26	9% more enjoyable	

Table 1.
(Continued)

Study	Auditory stimuli	Food/drink	DV	Study design	Sample size	% difference	Effect size (Cohen's *d*)
Fiegel *et al.*, 2014	4 genres (jazz, classical, hiphop, rock), single or multiple performers	Emotional (chocolate) *vs* non-emotional (bell pepper) food	VAS scale 15 cm: flavour intensity, pleasantness, texture liking, overall liking	Within participants (genre), between participants (single/multiple performers)	99		
Spence *et al.*, 2014c, study 1	White light, red light, green light + sour music, red light + sweet music	Red wine	7-point scales: fresh-fruity, intensity, liking	Within participants	1580		
Spence *et al.*, 2014c, study 2	White light, green light, red light + sweet music, green light + sour music	Red wine	7-point scales: fresh-fruity, intensity, liking	Within participants	1309		
Reinoso Carvalho *et al.*, 2015b	Sweet, bitter, medium soundtracks	Chocolate (bitter, medium, sweet)	9-point scale: bitter-sweet. 5-point scale: less–more bitter or less–more sweet	Within participants	24		

Table 1.
(Continued)

Study	Auditory stimuli	Food/drink	DV	Study design	Sample size	% difference	Effect size (Cohen's d)
Wang and Spence, 2015a	Classical music (Debussy, Rachmaninoff)	Wine (1 white and 1 red)	VAS scale 100 mm: wine–music match, fruitiness, acidity, tannins, richness, complexity, length, pleasantness	Between participants	64	15% more fruity, 42% more acidic	0.38 (fruitiness), 1.10 (acidity)
Reinoso Carvalho *et al.*, 2016e, experiment 1	Sweet, bitter sountracks	Belgian beer	7-point scales: sweet, bitter, sweet–bitter, strength, enjoyment	Within participants	113	20% sweeter (sweet scale), 16% (sweet–bitter scale)	0.40 (sweet), 0.41 (bittersweet)
Reinoso Carvalho *et al.*, 2016e, experiment 2	Sweet, sour soundtracks	Belgian beer	7–point scales: sweet, sour, sweet–sour, strength, enjoyment	Within participants	117	20% sweeter (sweet scale), 10% sweeter (sweet–sour scale), 22% more liked	0.42 (sweet), 0.28 (soursweet), 0.52 (liking)
Wang and Spence, 2016	Melodies with consonant and dissonant harmonies	Juice mixture	10-point scales: music liking, drink liking, sour-sweet scale	Within participants	39	19% sweeter	0.43

Table 1.
(Continued)

Study	Auditory stimuli	Food/drink	DV	Study design	Sample size	% difference	Effect size (Cohen's *d*)
Reinoso Carvalho *et al.*, 2017	Legato, staccato soundtracks	Chocolate	7-point scales: sweetness, bitterness, creaminess, liking, chocolate–music match, music liking	Within participants	116	11% creamier and sweeter, 8% less bitter	0.27 (creamy), 0.27 (sweet), 0.23 (bitter)
Wang *et al.*, 2017c, experiment 2	Spicy soundtrack, sweet soundtrack, white noise, silence	Salad	11-point scales for expected and actual ratings of: sweetness, spiciness, flavour intensity, liking	Between participants	180 (45 per cell)	30% spicier (expected, *versus* silent condition)	0.89
Wang *et al.*, 2017c, experiment 4	Spicy soundtrack, silence	Salsa, mild and medium spicy	11-point scales: flavour intensity, pleasantness, spiciness	Within participants	40	16% spicier	0.4
Wang and Spence, 2018	Melody with consonant and dissonant harmonies; images with happy/sad child	Juice mixture	11-point scales: sour–sweet, liking	Within participants	49	18% sweeter	0.28

Table 1.
(Continued)

Study	Auditory stimuli	Food/drink	DV	Study design	Sample size	% difference	Effect size (Cohen's *d*)
Hauck and Hecht, 2019	Classical music (Berg, Tchaikovsky)	Red wine, white wine, sugar water, citric acid solution	11-point scales: overall liking, sweet, sour, salty, bitter, foul, floral, aromatic, fruity, lively, gloomy, harmonic, light, zingy and refreshing, powerful and heavy, subtle and refined, mellow and soft	Within participants (misreported in paper as between participants!)	115	10% more liked	0.3
Höchenberger and Ohla, 2019, study 1	Sweet, bitter soundtracks, silence	Cinder toffee	0–100 VAS: bitter–sweet, pleasantness	Within participants	20	8% sweeter	0.55
Höchenberger and Ohla, 2019, study 2	Sweet, bitter soundtracks, silence	Cinder toffee	0–100 VAS: sweet, bitter, salty, sour, pleasantness	Within participants	20		
Wang et al., 2019	Sweet, bitter soundtracks, silence	Juice mixture	9-point scales: sweetness, bitterness, sourness, liking	Mixed (soundtrack, colour = within participants; aroma = between participants)	331 (~50 per cell)	8% sweeter (sweet vs bitter soundtrack), 4% sweeter (control vs bitter soundtrack)	0.27 (bitter vs sweet soundtrack), 0.16 (bitter vs control)

high pitch and sweet, and sour, taste descriptors, food names, as well as an association between low pitch and bitter food names (Crisinel and Spence, 2009, 2010a). That said, a potential confound here is that participants might have matched pitches to the linguistic features of the food names themselves, rather than the (imagined) tastes of the foods. Simner *et al.* (2010) demonstrated that phonetic features were reliably matched to basic tastes at two different concentrations, especially with sweet tastes being matched to lower values in terms of vowel height, vowel front/backness (where lower values correspond to more back in vowel space), and spectral balance compared to sour tastes (see also Motoki *et al.*, 2018).

In order to make sure that participants were matching sounds to imagined food tastes rather than to linguistic features of the food names, Crisinel and Spence (2010b) conducted another study using actual taste and aroma solutions. In this case, the participants had to match each taste sample to a musical note (one of 13 notes from C2 to C6, in intervals of two tones) and a class of musical instruments (piano, strings, winds, and brass). The results demonstrated that for a number of these tastes and aromas, the participants were consistent in terms of the notes and instruments that they felt went especially well together. So, for instance, sweet and sour tastes were mapped to higher-pitched sounds, while bitter tastes were mapped to lower-pitched sounds. In addition, sweet tastes were mapped to piano sounds whereas bitter and sour tastes were mapped to brass instruments. In terms of aromas, fruity notes such as apricot, blackberry, and raspberry were all matched with higher (rather than lower) musical notes, and with the sounds of the piano and often also woodwind instruments, rather than with brass or string instruments. By contrast, lower-pitched musical notes were associated with musky, woody, dark chocolate, and smoky aromas, bitter tastes, and brassy instruments instead (see also Crisinel and Spence, 2012a, for an extensive exploration of wine odour–musical note matching; and Burzynska, 2018, for practical explorations in this space).

Approaching the sound-taste correspondence problem from a somewhat different angle, Mesz *et al.* (2011) had nine professional musicians improvise freely on the theme of basic taste words (bitter, sweet, sour, and salty). The resulting improvisations were analysed, revealing consistent musical patterns for each taste. Specifically, bitter improvisations were low-pitched and legato, salty improvisations were staccato, sour improvisations were high-pitched and dissonant, and sweet improvisations were consonant, slow, and soft. A follow-up experiment had 57 non-musicians choosing a basic taste word that best matched a subset of the improvisations. The participants performed significantly better than chance (around 68% correct, as compared to chance level of 25%; see Mesz *et al.*, 2012). Similarly, Knoeferle *et al.* (2015) reported on a study in which regular participants matched auditory properties (pitch height, roughness, sharpness, discontinuity, tempo, sharpness, and attack) to basic taste words (sweet, sour, salty, and bitter) by using a series of sliders to control the auditory properties of a short chord progression. More recently, Guetta and Loui (2017) created violin soundtracks consisting of the same melody played in four different styles that were informed by previous studies on basic taste and music associations. The participants in this study were shown to reliably match auditory clips to taste words (sweet, sour, bitter, salty) at above chance levels, as well as matching the auditory clips to custom-made chocolates expressing the same basic tastes.

In an overarching survey of taste-corresponding soundtracks, Wang, Woods and Spence (2015) conducted an online study in which 100 participants listened to samples from 24 soundtracks and chose the taste (sweet, sour, salty, bitter) that best matched each sample. Overall, sweet soundtracks tended to have the most consensual response (participants chose sweet 56.9% of the time for sweet soundtracks, compared to 25% random chance), whereas bitter soundtracks were the least effective (participants chose bitter 31.4% of the time for bitter soundtracks). Moreover, a follow-up study demonstrated that associations between soundtracks and tastes were partly mediated by pleasantness for sweet and bitter tastes, and emotional arousal for sour tastes. Over the last few years, researchers have also started to explore the crossmodal correspondences that link to a number of more complex gustatory qualities such as spicy (Wang, Keller and Spence, 2017c), creamy (Reinoso Carvalho *et al.*, 2017), and oak (e.g., in a wine; Wang *et al.*, in press). Other food-and-beverage qualities that are potentially relevant that have now been rendered in auditory form include temperature (see Wang and Spence, 2017b; see also Septianto, 2016) and even wine styles (Spence *et al.*, 2013; Wang and Spence, 2015a, 2017a; see Spence and Wang, 2015c, for a review).

One other crossmodal correspondence that has not, as yet, received much empirical interest is the sound/taste correspondence that is based on perceived intensity. Wang *et al.* (2016), for instance, gave people solutions containing one of the five basic tastes at one of three different stimulus intensities. The results revealed that participants chose louder sounds to match the more intense tastes. Elsewhere, it has been noted that when the music or soundscape is presented while people are tasting, the latter's ratings of taste intensity tend to be higher than when tasting in silence instead (though note here that different results may be obtained if what is heard is classified as noise; e.g., see Yan and Dando, 2015).

As has been noted already, beyond a subjective feeling that certain auditory stimuli match a particular corresponding taste quality, such correspondences have also been documented using Implicit Association Test (IAT)-type tasks (Crisinel and Spence, 2009, 2010b). More recently, Padulo *et al.* (2018) went

on to demonstrate that the speed with which participants ($N = 86$ participants) classified food images as either salty or sweet was facilitated by playing
the matching rather than mismatching music, neutral environmental sounds,
or else when performing the task in silence. The participants in this study
were significantly faster to classify images as salty when accompanied by a
'salty' sound than by a 'sweet' sound, neutral environmental sound (that in
pre-testing was equally matched with each taste), or silence. Finally, here, beyond the effect of sonic seasoning on the consumers' tasting experience, there
is also some preliminary evidence to suggest that the music playing in the
background might also influence the way in which those in the kitchen, or bar,
season the food and drink they prepare (Kontukoski *et al.*, 2015; see also Liew
et al., 2018).

North's (2012) results (reported in section 3; see also Silva, 2018), might
strike some readers as providing an example of 'sonic seasoning'. That said,
Spence and Deroy (2013a) argued that crossmodal correspondences between
basic sensory features of musical (or auditory) stimuli should perhaps be distinguished from the emotional attributes, or connotation, that may be associated with a piece of music. The latter may perhaps influence people as a result
of priming, without there necessarily being any natural affinity between the
stimuli concerned. However, the distinction is by no means cut-and-dried, and
may benefit from further consideration of the similarities and differences between these two kinds of crossmodal influence. The waters become especially
muddy, here, once one recognizes the growing interest amongst researchers in
those crossmodal correspondences that appear to be mediated, at least in part,
by the affective/emotional valence of the component stimuli.

5.1. When Crossmodal Correspondence Becomes 'Sonic Seasoning'

In terms of research on the crossmodal correspondences between sonic properties and gustatory/olfactory attributes, it is important to stress that the mere
existence of a crossmodal correspondence (Note 12) does not in-and-of-itself
guarantee that playing the corresponding tone, soundscape, or musical excerpt will necessarily always modulate the taste/flavour (Knöferle and Spence,
2012). In order for such crossmodal effects on perception (or, at the very least,
on people's ratings) to be observed, it would appear that certain conditions (or
constraints) need to be met. Figure 1 addresses some of the potential mechanisms with which sonic seasoning soundtracks can give rise to perceptual (or
evaluated) differences. Wang's PhD thesis work (Wang, 2017) found evidence
to support the notion that sound can change food evaluation *via* the mechanisms of sensory expectations, attention capture, and emotion mediation.

One cannot simply turn water into wine by picking the right musical accompaniment. Rather, it would seem likely that the taste/aroma/flavour must be
present in the food or beverage stimulus to begin with in order for the taster's

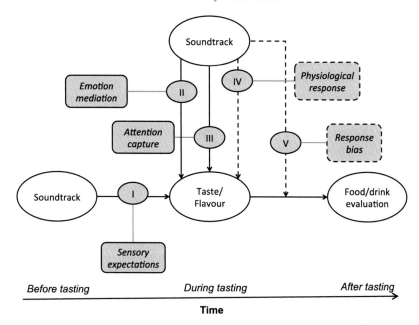

Figure 1. Schematic diagram summarizing the various ways in which sonic seasoning might influence tasting/flavour evaluation at different points in time, from Wang's Oxford University DPhil Thesis (Wang, 2017). Dashed lines denote mechanisms for which no evidence was found in research so far, whereas continuous lines denote those mechanisms garnering empirical support. For relevant studies, please see: Mechanism I — sensory expectations (Wang *et al.*, 2017c), Mechanism II — emotion mediation (Wang and Spence, 2018; Wang *et al.*, 2016), Mechanism III — attentional capture (Wang *et al.*, 2017a, b), Mechanism IV — physiological response (Wang *et al.*, 2017d), Mechanism V — response bias (see Wang, 2017, Ch. 4) (Note 13).

experience of that attribute to be modified auditorily. Although no one knows for sure, what we suspect may be happening is that sound draws the taster's attention to something in their experience, and by so doing, it makes that element more salient (see Spence, 2014b; Wang, 2017, Chapter 6; cf. Klapetek *et al.*, 2012). At the same time, however, by drawing a taster's limited attentional resources away from other elements in their experience, the latter are likely to become less salient components of the tasting experience. As such, our suspicion is that those multisensory tasting experiences that are more complex to begin with, in the sense of more flavours being present in the tasting experience (see Spence and Wang, 2018a, for a review, of the various meanings of complexity as far as the chemical senses are concerned), may present more opportunity for selective attention to be drawn crossmodally (and presumably also exogenously; see Spence, 2014b) to one element in the experience, if compared to when a tasting experience presents only a unitary dimension to begin with.

It could also be imagined that sonic seasoning might be more effective under those conditions in which the taster is unfamiliar with exactly what they are tasting. Otherwise, should an easily recognized branded product like Coca-Cola be presented, say, then the taster might perhaps rely more on their memory of the taste/flavour, than on their actual tasting experience (though, that said, see McClure *et al.*, 2004, for evidence that branding effects work even with familiar brands of cola). Look carefully, and you will see that we often present unusual mixtures of fruit juice, or else serve wines blind, for just this reason (e.g., Wang and Spence, 2015a, 2016, 2017c). Indeed, elsewhere in the field of audiovisual research, there have been frequent demonstrations that expectations have a bigger influence on our sensory processing when the input stimuli are weak, noisy, and/or ambiguous (de Lange *et al.*, 2018).

Furthermore, it is also important to note that low pitch, for instance (as but one example of an auditory feature), does not only correspond to a bitter-tasting food or beverage product. Rather, it corresponds to a whole host of other attributes in a variety of senses (see Parise, 2016; Spence, 2011). Note that we usually ask our participants to estimate specific tastes and by so doing presumably draw their attention to that particular element in the tasting experience. Indeed, it is easy to imagine how the taste-relevant correspondence somehow needs to be made salient to the taster (cf. Schietecat *et al.*, 2018). Otherwise, there might be a danger of the taster concentrating on the loudness of the sound or perhaps its duration instead, rather than necessarily on the relevant dimension, in this case, namely, the pitch. Crossmodal correspondences, in other words, are typically not established automatically (e.g., Getz and Kubovy, 2018; Spence and Deroy, 2013b). In this regard, it is interesting to note that when the culinary artist Caroline Hobkinson served the bitter-sweet sonic cake pop at her pop-up dining experience at the House of Wolf restaurant, diners were actually encouraged to take out their phone and dial one number in order to listen to 'sweet' music while dialling another number if they wanted to bring out the bitterness in their dessert instead (see Spence, 2017b).

The fact that people may be able to choose which music they think best matches with different available food choices prior to a sonic seasoning task, say, could have further implications in the overall multisensory tasting experience as well. For instance, in Reinoso *et al.*'s (2015b) study, three soundtracks were produced (one sweet, one bitter, and one in-between). The results revealed that what people heard exerted a significant influence over their taste ratings of three available types of chocolate. However, when the results were analysed on the basis of the participants' individual music–chocolate matches (rather than the average response of the whole group of participants), somewhat more robust crossmodal effects were revealed.

There are also two further points that are perhaps worth mentioning here. One might well reasonably wonder whether sonic seasoning would work better when sounds are presented over headphones, so in some sense leading to the sound being located in the same location (i.e., inside the head) where the taste is experienced as originating from (Spence, 2016a). While we are not aware of anyone having tested this experimentally as yet, research from elsewhere in the world of multisensory perception clearly shows that spatial colocation (i.e., in the sense of sounds coming from headphones *vs* external loudspeakers) can sometimes modulate the magnitude of any crossmodal effects that are reported (Di Luca *et al.*, 2009; Soto-Faraco *et al.*, 2002; Spence and Driver, 1997; though see Kaiser *et al.*, 2016). At the same time, however, the very act of wearing headphones may perhaps lead participants to focus their attention toward their ears (and hearing), which could also enhance any influences of sound on the eating experience. Potentially relevant here, therefore, it is worth noting that Crisinel *et al.* (2012) used headphones to present the bitter and sweet soundscapes, whereas Höchenberger and Ohla (2019), in their attempt to replicate Crisinel *et al.*'s results, actually switched to presenting the sounds from external computer loudspeakers instead. Now, this may not turn out to matter much. Nevertheless, it is probably a factor that should be borne in mind (and, one presumes, noted by the researchers concerned).

The second point to bear in mind here is that crossmodal influences of audition on tasting are often quite subtle — showing up more often at the group level rather than necessarily as a striking change at an individual level (though the latter does, sometimes, occur). This may be attributable to the fact that we have an 'assumption of unity' concerning food and drink (see Woods *et al.*, 2010). Namely, we expect most food and beverage products to taste the same from start to end (see Note 14). As such, if people are aware that what they are tasting, or have very good reason to believe that what they are tasting, is the same, the unity assumption may well prove more powerful than the crossmodal effect of audition. In this regard, sonic seasoning is quite different from something like the McGurk effect, where the illusion is so powerful that observers mostly cannot override it at will (see Note 15).

Meanwhile, in terms of neural changes seen as a consequence of playing crossmodally corresponding music while tasting, some exciting preliminary neuroimaging results have recently started to appear (see Callan *et al.*, 2018). Given that sound has been shown to alter people's sensory expectations, we may expect to find some neurological evidence that is relevant. For instance, human neuroimaging and animal electrophysiology has shown that expectations (in terms of audiovisual studies, at least) can modulate sensory processing at both early and late stages of information processing, and the response modulation can be either dampened or enhanced depending on the context (see de Lange *et al.*, 2018; Piqueras-Fiszman and Spence, 2015, for reviews). Similar expectancy effects have also been shown when participants are informed that a drink will have a specific taste. Namely, participants who are told to expect a very sweet drink when given a less sweet drink showed greater taste cortex activation, as compared to those who received the same drink without this expectation (Woods *et al.*, 2011b; see Spence, 2016b, for a review; see also Geliebter *et al.*, 2013). Finally, Wang *et al.* (2017d) investigated a possible direct physiological effect of crossmodally corresponding music by measuring the rate of salivation while participants listened to a sour soundtrack, watched a muted video of a man eating a lemon, or else sat in silence. While the salivation rate was significantly higher during the lemon video condition than the silent baseline condition, no such difference was observed between the sour soundtrack condition and baseline condition.

5.2. *Individual Differences*

One question that often crops up is whether such crossmodal effects between sound properties and taste are the same in different cultures (of course, a similar question might well crop up with regard to the different music styles discussed in section 3.3). While a thorough analysis has yet to be conducted, Knoferle *et al.* (2015) were at least able to demonstrate that four variations on a musical theme that had been designed to match each of the four basic tastes (e.g., sweet, sour, bitter, and salty) gave rise to almost as high agreement (or concordance/consensuality) about the matching, or corresponding, taste in a population from India as in a group from North America (note that, in this case, the compositions themselves had been generated in Germany).

Another individual difference here relates to genetic differences in terms of supertaster status. This has also been demonstrated to play a role in terms of sonic seasoning effects. For instance, using a mixed model design, Wang had 27 participants taste 70% and 85% cacao chocolate while listening to sweet and bitter soundtracks (Wang, 2017, Ch. 9). All participants then took a PTC taste strip test at the end of the study. The results revealed an intriguing split when it came to the influence of music. While there were no differences between the two taste sensitivity groups for 70% chocolate, when it came to the more bitter 85% chocolate, the high taste sensitivity group appeared to be more influenced by the different soundtracks than the low sensitivity group (i.e., they found a bigger difference in the taste of the 85% chocolate between the bitter and sweet soundtrack; cf. Crisinel and Spence, 2012b).

Another question relates to the role of expertise, both in terms of musical expertise and in terms of taster expertise. In Wang *et al.*'s (2015) study, where 24 excerpts from soundtracks were tested in terms of their taste associations, musical expertise was found to influence how participants made their sound–taste correspondences for one of the soundtracks. *Makea*, composed by musician and researcher Bruno Mesz, was a soundtrack featuring high-pitched piano instrumentation and dissonant chords putatively associated with sweetness. Results from testing 100 participants turned out to be subtler: those with no musical background were significantly more likely to match the soundtrack with sweetness than those with musical experience, for whom bitterness was the most common choice. This was probably due to the fact that musical novices tend to focus on timbre whereas experts tend to focus on melody and harmony instead (Wolpert, 1990). Therefore, perhaps the novices matched the high-pitched piano sounds to sweetness, while the more experienced listeners matched the dissonant chords to bitterness.

While there has not yet been a direct comparison between expert tasters with regular consumers, it has recently been demonstrated that even wine expert's judgments of the properties of wine could be influenced by the music playing in the background. In particular, Wang and Spence (2017c) tested 154 wine professionals attending the International Cool Climate Wine Symposium in two studies. Their first study replicated previously demonstrated effects of sweet and sour soundtracks, where participants rated an off-dry white wine as sweeter and less sour (on two independent scales), when they tasted while listening to the sweet soundtrack compared to the sour soundtrack. In a second study, the participants tasted a pair of chardonnays and evaluated wine-specific terminology (length, balance, body) while listening to two soundtracks with contrasting auditory textures (sparse *vs* full). Both wines tasted while listening to the sparser soundtrack were associated with fuller body, better balance, and longer length, compared to the soundtrack with fuller texture (see also Burzynska, 2018). The amount of wine tasting experience (in terms of years) did not moderate the influence of music on the participants' sensory wine evaluation (see Note 16).

5.3. Tell Me About the Taste of the Product vs *Tell Me About Your Tasting Experience*

In many of the experiments that have been conducted to date on the topic of sonic seasoning, the participants have deliberately been given the impression that they are actually (or might well be) tasting a range of different food stimuli, or else used mixed-design models in which each participant gets to tasting multiple different foods (e.g., Reinoso Carvalho *et al.*, 2015b; Wang and Spence, 2015a; Wang *et al.*, 2017c). Contrast this with the situation in Höchenberger and Ohla's (2019) recent study in which, from the way in which the materials and method are described, the participants were simply presented with a tray of pieces of cinder toffee. Given this arrangement, where the participants were free to pick any piece on each of the 27 trials, one could presumably safely infer that the stimuli must be the same. As such, there arises an important distinction here, between two similar sounding judgments. If participants report on the taste/flavour of the chocolate, their response might be dissociated from how they actually subjectively experience the taste/flavour of the chocolate.

By analogy, imagine the different responses that you would be tempted to give if you just saw the lighting strike a long way off on the horizon, and then three seconds later heard the crack of the thunder. If asked what just happened, you will say that there was a single bolt of lightning (with simultaneous visual and auditory properties). However, if asked what you just perceived, then you would, we imagine, come out with a different answer, namely that you first saw the lightning strike, and a few seconds later you heard the crack of the thunder (Spence and Squire, 2003). Notice how, in this case, you are able to dissociate your knowledge of what is out there from your perception of the event, given your priors and beliefs about the world.

At the same time, however, there is a growing realization that certain food and beverage products have a temporally-evolving flavour profile (Wang, Mesz and Spence, 2017a, b), and hence synchronizing the musical properties to the evolving attributes of the tasting experience becomes an increasingly important issue. Evidence from elsewhere in the field of multisensory research would appear to suggest that temporally synchronized soundscapes are likely to have a more pronounced influence over the tasting experience than when food is tasted at random points in the music (though see Houge and Friedrichs, 2013, for a discussion of the difficulty of synchronising music with food in a restaurant setting; see also Rozin and Rozin, 2018). Now, of course, all these caveats likely mean that while 'sonic seasoning' has an important role in multisensory experience design (see Spence, 2019), there may be less that is directly applicable from a marketing perspective (or rather the application might be more on the advertising side than on the choice of music to play in-store/restaurant).

6. Conclusions

As this review of the rapidly-expanding literature documenting crossmodal contributions of audition to food perception and consumer behaviour has hopefully made clear, product-extrinsic sounds exert a profound influence over various aspects of people's perception of the aroma, taste, and flavour of a wide variety of food and drink items. The sonic properties of the ambient soundscape also exert often-unacknowledged effects on consumer behaviour across a wide variety of food-related contexts (e.g., see North *et al.*, 1997, 1999; Zellner *et al.*, 2017). Importantly, while many of these effects have been studied on participants in the laboratory, they have also been documented in customers in a number of more ecologically-valid settings too, such as restaurants, shops, bars, cultural institutions, and wine bars. It is perhaps because these sounds are mostly unrelated to the food or drink itself in these studies, people rarely seem to be aware of just how much influence music/noise can have over what they taste, and how much they enjoy the experience.

6.1. Neuroscientific Explanations of the Auditory Influence on Food Perception and Consumer Behaviour

In the future, the results of neuroimaging research will likely also help to confirm whether we are indeed looking at four distinct routes (or mechanisms) underlying the crossmodal influence of auditory on food perception and consumer behaviour outlined here (see Callan *et al.*, 2018, for some intriguing preliminary data). Alternatively, however, we should perhaps also remain open to the possibility that despite the background literature (for these four categories; namely, background noise, background music, sensation transference, and crossmodal correspondences) being so separate, some meaningful consolidation can take place, either between these seemingly distinct areas of research, or at the very least, at their boundaries.

As yet, while the behavioural/psychophysical data documenting the influence of what we hear on what we taste continues to build up, our cognitive neuroscience understanding of the neural mechanism(s) underlying such crossmodal effects continues to lag far behind. To the extent that somewhat different physiological/neurophysiological mechanisms do underlie each of the identified routes by which what we hear influences what we taste and smell, then one might reasonable expect somewhat different networks of neural activity to be involved. Here it is perhaps interesting to note that while direct cortical connections between olfactory and auditory brain areas were discovered in the rat a few years ago (Wesson and Wilson, 2010, 2011), leading one excitable commentator to introduce the new term 'smound', for the combination of smell and sound (see Peeples, 2010; see also Cohen *et al.*, 2011), their role and even the question of whether similar connections also exist in humans has not been addressed as yet, at least as far as we are aware. Moving forward, of course, having a better cognitive neuroscience understanding of what is going on in the brain while people taste, purchase, and/or consume food and drink while different kinds of music or noise are present will likely help further our understanding in this area.

6.2. Product-Extrinsic Multisensory Contributions to Food Perception and Consumer Behaviour

What is also worth noting is that all of the studies that have been reviewed here have manipulated only a single sense at a time, namely audition. However, in the real world, what we hear is clearly going to be but one element of the total multisensory atmosphere. The visual, olfactory, and tactile attributes of the atmosphere clearly also matter, and likely interact with the auditory soundscape in the taster's experience (see Spence, 2017b, for a recent review). Hence, researchers are now starting to assess how, for example, the visual attributes of the environment, combined with the auditory atmosphere, can influence a consumer's behaviour (e.g., Sester *et al.*, 2013; Spence *et al.*, 2014b, c; Wang and Spence, 2015b; Wang *et al.*, 2019; Wansink and Van Ittersum, 2012). Researchers have also started to assess different ways to effectively present music as part of a food/drink product's identity. This is being explored by means of semantically framing the music that is presented while tasting (i.e., by presenting the music as the main source of inspiration of a food/drink product's formula; and/or by including such music as part of a product's presentation — as in a kind of multisensory packaging; see Reinoso Carvalho *et al.*, 2015a, 2016c). This, though, undoubtedly adds to the complexity of the problem under study.

6.3. *Multisensory Experience Design*

Given the growing literature on music and soundscape's influence on the multisensory tasting experience, there is a growing interest in using technology to synchronize aspects of the auditory stimulation with the tasting experience (Reinoso Carvalho *et al.*, 2016a; Velasco *et al.*, 2016; see Spence, 2019, for a review). This is undoubtedly a rich area for creative practice. The Chocolate Symphony presented at the 2018 IMRF meeting in Toronto is a very recent example (see http://maximegoulet.com/symphonic-chocolates/). The city of Brussels (Belgium) also recently-funded a project entitled 'The Sound of Chocolate' (www.thesoundofchocolate.be), where chocolate boxes were sold alongside music that was designed to enhance certain aspects of these chocolate's taste and flavour.

In fact, in some cases specially composed atmospheric soundscapes or specially chosen pieces of music are now being developed to complement the dishes served on the ground (see Spence and Youssef, 2016; Spence *et al.*, 2011), and even in the air (FinnAir (Note 17); British Airways: Victor, 2014). A number of food and beverage brands have also started to capitalize on the opportunities provided by connecting their product offering with specific pieces of music (e.g., though sensory apps; see Spence, 2019, for a review). There is, though, at the same time a question, at least amongst some, of 'why bother?' (see Spence and Wang, 2015d, for a review of those who have taken such a position). Actually, it is here that the effort to reduce sugar intake *via* sound, and/or colour, by let's say using 'smart' technologically-enhanced cups (reported by Blecken, 2017), not to mention the latest pitch-overeating effects that have been demonstrated by Lowe *et al.* (2018), becomes so relevant.

The latter researchers just reported a study that capitalized on pitch/size cross-modal associations in order to evaluate whether sounds of different pitches would lead to different serving sizes. As the authors predicted, lower-pitched ads led to larger serving sizes as compared to higher-pitched ads (see also Lowe and Haws, 2017).

6.4. *Implications for Public Health*

A case can be made that the loud, fast music so often piped-out at restaurants and bars may be exerting a negative effect over consumer perception and behaviour. As such, some have suggested that there may be important — if largely unacknowledged — consequences of the soundscapes in which we come into contact with food and drink products (all the way from the shopping until the tasting process; Keller and Spence, 2017; Liu *et al.*, 2018; Mamalaki *et al.*, 2017). Here it is worth noting that long-term exposure to transportation noise has been linked to obesity, and that combined exposure to different sources of noise has been shown to be particularly harmful (e.g., see Pyko *et al.*, 2017). One can make an analogy with the multiple sources of background noise in a Sports Bar, say, where music, background conversation, and the game showing on the screens all compete in an auditory cacophony. As far as we are aware, the question of the relevance/impact of the number of sources of noise/music in the environments in which we eat and/or drink has yet to be investigated. However, attention is already starting to turn to the impact that loud background noise may be having on children's fruit and vegetable consumption in the school canteen (Graziose *et al.*, 2019).

On the flip side, however, it is presumably only by recognising the effect of the ambient soundscape on tasting that we will be in a better position to design those soundscapes that may have a better chance of promoting, let's say, healthy eating (see Blecken, 2017; Ragneskog *et al.*, 1996), or food shopping behaviour in all who hear them (see Spence, 2012). For instance, Bueno and Bergamasco (2018) recently assessed the effect that the association of taste and music had on the mood of Brazilian children. In this study, children tasted sweet and bitter solutions in silence *vs* listening to music that was pre-selected as evoking either happiness or sadness. When the children were tasting bitter food and listening to happy music, their emotional state switched from sad to happy. A similar effect was found when they were tasting sweet food and then listened to sad music, where their emotional state switched from happy to sad (cf. Reinoso-Carvalho *et al.*, 2019). Hence, as a case in point, consider only the school lunch cafeteria or work canteen, where strategically playing the right sort of background music, or soundscape (whatever that might be) might encourage consumers to choose more vegetables or sustainably-sourced protein (here one need only think of Zellner *et al.*'s, 2017, study with Spanish *vs* Italian meals served in the student cafeteria). Sonic seasoning might also play a role at the condiment station, where a sweet background track might just induce people to add less sugar to their coffee (see Blecken, 2017; Lowe *et al.*, 2018). That said, long-term follow-up studies are urgently needed in order to ascertain whether these sonic influences last into the longer-term (i.e., do they persist beyond the span of an individual laboratory experiment).

Notes

1. Emile Peynaud, a famous French oenologist, hinted at the distracting effect of noise when he stated that: *"The sense of hearing can interfere with the other senses during tasting and quiet has always been considered necessary for a taster's concentration. Without insisting on absolute silence, difficult to obtain within a group in any case, one should avoid too high a level of background noise as well as occasional noises which can divert the taster's attention."* (Peynaud, 1987, p. 104).

2. It is worth noting here that the latest evidence suggests that people's response to umami differs by culture/country (see Cecchini *et al.*, 2019).

3. In this regard, one might speculatively want to consider airplane noise as a kind of 'sonic seasoning' (see section 5). However, it is as yet unclear whether consumers consider airplane noise a particularly good match for the taste of umami, as would be needed if one wanted to establish the crossmodal correspondence underpinning this particular crossmodal effect.

4. However, while such an explanation may sound promising, it is perhaps worth noting that not everyone necessarily believes in the possibility of crossmodal masking; see McFadden *et al.*, 1971).

5. Note that the loud music is presumably also congruent with the brand, and this may be perceived positively as a result.

6. All of this, while at the same time performing a shadowing task involving listening to and repeating a news story. Pellegrino *et al.* (2015) have also concluded that conversing is a preferred activity in eating atmospheres (see also Lindborg, 2016), although it can alter the consumer's ability to discriminate basic differences between foods or beverages. These results also suggest that the judgment of the flavour of foods that give rise to high levels of mastication sound tend to be less susceptible to the influence of background noise.

7. Here, one might even consider recent findings that have shown that Pavlovian conditioning can give rise to hallucinations (Powers *et al.*, 2017).

While, to date, such hallucinations have only been studied in the audio-visual domain, there would seem no good reason, *a priori*, as to why such perceptually vivid hallucinations (or vivid sensory mental imagery) wouldn't also extend to the chemical senses as well (see also Spence and Wang, 2018a, on the topic of imagined flavours complementing directly perceived flavours).

8. In recent years, it has become increasingly easy to capture big data concerning people's eating behaviours *via*, for instance, smartphones. Nowadays, most smartphones have a microphone capable of measuring ambient noise levels, and a platform for recording one's food habits, not to mention Instagramming the dishes that one has chosen/eaten (e.g., see Ofli *et al.*, 2017). Especially relevant here, 'Soundprint', offers the opportunity for the crowd-sourced measurement of noise levels in restaurants. Analysis of such data, collected using the novel SoundPrint smartphone app, has already started to reveal a number of intriguing findings, such as the fact that the average noise level recorded in more than 2250 restaurants and bars in New York City, was 78 dBA in restaurants and 81 dBA in bars. Note that such sound levels do not allow ready conversation and may pose a danger for noise-induced hearing loss and other non-auditory health issues (Fink, 2017). Worryingly, managers were also found to underestimate the actual sound levels in their venues (Farber and Wang, 2017).

9. Note that the valence of the music had been established by Reinoso Carvalho *et al.* (2019) in their study, by having the participants evaluate each song using the positive and negative affect schedule (PANAS).

10. Alternatively, however, it might be argued that 'sensation transference' is a qualitatively different phenomenon that the semantic priming that was discussed in the preceding section.

11. Are they, for instance, based on the statistics of the environment (Ernst, 2007; Spence, 2011), or perhaps reflect some sort of innately determined correspondence? Or are they the product of transitive properties (e.g., bitterness corresponds with low pitch because both correspond with dark colours or negative emotion; see Palmer *et al.*, 2013)?

12. Defined as a 'feeling of rightness' that certain sound properties match, or go together well with specific taste properties; i.e., that bitter tastes seem to match low-pitched soundscapes, or pieces of music.

13. One interesting consideration here is the extent to which the influences outlined in Fig. 1 in the case of 'sonic seasoning' could also be applied to the case of the influence of background music, or even background noise, on tasting covered in sections 2 and 3.

14. Though drinks like quality wine are interestingly different in this regard, possibly due to their complex nature (see Wang *et al.*, 2017a, b).

15. Though it is perhaps worth noting here that the recent history of congruent *vs* incongruent stimuli (presumably affecting the priors we hold about the likelihood that what we see and hear belong to the same speech event) has even been shown to modulate the magnitude of the McGurk effect (Gau and Noppeney, 2016; Nahorna *et al.*, 2012, 2015), one of the classic examples of multisensory perception. The strength and robustness of even the most reliable of multisensory illusions, or crossmodal effects, in other words, may also be subject to our beliefs about the causal structure of the world around us. See Delle Monache *et al.* (2008) on the use of incongruent sounds at a dining experience.

16. While the focus here is on tasting, it is worth noting that there is also a long history of researchers assessing the crossmodal correspondences between food-relevant odours and musical notes too (see Bronner *et al.*, 2012; Crisinel and Spence, 2012a; Deroy *et al.*, 2013; Piesse, 1891).

17. Retrieved from https://www.finnair.com/cn/gb/stevenliu/en (August, 2018).

References

Areni, C. S. and Kim, D. (1993). The influence of background music on shopping behavior: classical versus top-forty music in a wine store, *Adv. Consum. Res.* **20**, 336–340.

Bach, P. J. and Schaefer, J. M. (1979). The tempo of country music and the rate of drinking in bars, *J. Stud. Alcohol* **40**, 1058–1059.

Belluz, J. (2018). Why restaurants became so loud — and how to fight back. "I can't hear you.", *Vox*, April 18th. https://www.vox.com/2018/4/18/17168504/restaurants-noise-levels-loud-decibels.

Biswas, D., Lund, K. and Szocs, C. (2019). Sounds like a healthy retail atmospheric strategy: effects of ambient music and background noise on food sales, *J. Acad. Market. Sci.* **47**, 37–55.

Blecken, D. (2017). Hold the sugar: a Chinese café brand is offering audio sweeteners, *Campaign*, February 13th. https://www.campaignasia.com/video/hold-the-sugar-a-chinese-cafe-brand-is-offering-audio-sweeteners/433757.

Bronner, K., Frieler, K., Bruhn, H., Hirt, R. and Piper, D. (2012). What is the sound of citrus? Research on the correspondences between the perception of sound and flavour, in: *Proceedings of the 12th International Conference of Music Perception and Cognition and the 8th Triennial Conference of the European Society for the Cognitive Sciences of Music*, Thessaloniki, Greece, pp. 142–148. Downloaded from http://icmpc-escom2012.web.auth.gr/files/icmpc-escom2012_book_of_abstracts.pdf.

Buckley, C. (2012). Working or playing indoors, New Yorkers face an unabated roar, *The New York Times*, July 19th. http://www.nytimes.com/2012/07/20/nyregion/in-new-york-city-indoor-noise-goes-unabated.html?_r=0.

Bueno, V. F. and Bergamasco, N. H. P. (2008). Effect of the association of taste and music on the mood of children, *Estudos de Psicologia (Campinas)* **25**, 385–393.

Burzynska, J. (2018). Assessing oenosthesia: blending wine and sound, *Int. J. Food Des.* **3**, 83–101.

Caldwell, C. and Hibbert, S. A. (2002). The influence of music tempo and musical preference on restaurant patrons' behavior, *Psychol. Mark.* **19**, 895–917.

Callan, A., Callan, D. and Ando, H. (2018). Differential effects of music and pictures on taste perception — an fMRI study, in: *Annual Meeting of the International Multisensory Research Forum*, Toronto, CA.

Cecchini, M. P., Knaapila, A., Hoffmann, E., Boschi, F., Hummel, T. and Iannilli, E. (2019). A cross-cultural survey of umami familiarity in European countries, *Food Qual. Prefer.* **74**, 172–178.

Clark, C. C. and Lawless, H. T. (1994). Limiting response alternatives in time–intensity scaling: an examination of the halo–dumping effect, *Chem. Senses* **19**, 583–594.

Clynes, T. (2012). A restaurant with adjustable acoustics, *Pop. Sci.* August. http://www.popsci.com/technology/article/2012-08/restaurant-adjustable-acoustics.

Cohen, L., Rothschild, G. and Mizrahi, A. (2011). Multisensory integration of natural odors and sounds in the auditory cortex, *Neuron* **72**, 357–369.

Crisinel, A.-S. and Spence, C. (2009). Implicit association between basic tastes and pitch, *Neurosci. Lett.* **464**, 39–42.

Crisinel, A.-S. and Spence, C. (2010a). A sweet sound? Exploring implicit associations between basic tastes and pitch, *Perception* **39**, 417–425.

Crisinel, A.-S. and Spence, C. (2010b). As bitter as a trombone: synesthetic correspondences in non-synesthetes between tastes/flavors and musical notes, *Atten. Percept. Psychophys.* **72**, 1994–2002.

Crisinel, A.-S. and Spence, C. (2012a). A fruity note: crossmodal associations between odors and musical notes, *Chem. Senses* **37**, 151–158.

Crisinel, A.-S. and Spence, C. (2012b). The impact of pleasantness ratings on crossmodal associations between food samples and musical notes, *Food Qual. Prefer.* **24**, 136–140.

Crisinel, A.-S., Cosser, S., King, S., Jones, R., Petrie, J. and Spence, C. (2012). A bittersweet symphony: systematically modulating the taste of food by changing the sonic properties of the soundtrack playing in the background, *Food Qual. Prefer.* **24**, 201–204.

Crocker, E. C. (1950). The technology of flavors and odors, *Confectioner* **34**, 7–8, 36–37.

de Lange, F. P., Heilbron, M. and Kok, P. (2018). How do expectations shape perception?, *Trends Cogn. Sci.* **1811**, 1–16. DOI:10.1016/j.tics.2018.06.002.

Delle Monache, S., Papetti, S., Polotti, P. and Rocchesso, D. (2008). Gamelunch: Forging a dining experience through sound, in: *CHI 2008 Extended Abstracts on Human Factors in Computing Systems*, pp. 2281–2286. ACM, Florence.

Demoulin, N. T. M. (2011). Music congruency in a service setting: the mediating role of emotional and cognitive responses, *J. Retail. Consum. Serv.* **18**, 10–18.

Deroy, O. and Spence, C. (2013). Weakening the case for 'weak synaesthesia': why crossmodal correspondences are not synaesthetic, *Psychon. Bull. Rev.* **20**, 643–664.

Deroy, O., Crisinel, A.-S. and Spence, C. (2013). Crossmodal correspondences between odors and contingent features: odors, musical notes, and geometrical shapes, *Psychon. Bull. Rev.* **20**, 878–896.

Di Luca, M., Machulla, T.-K. and Ernst, M. O. (2009). Recalibration of multisensory simultaneity: cross-modal transfer coincides with a change in perceptual latency, *J. Vis.* **9**, 7. DOI:10.1167/9.12.7.

Ernst, M. O. (2007). Learning to integrate arbitrary signals from vision and touch, *J. Vis.* **7**(7), 1–14. DOI:10.1167/7.5.7.

Farber, G. and Wang, L. M. (2017). Analyses of crowd-sourced sound levels, logged from more than 2250 restaurants and bars in New York City, *J. Acoust. Soc. Am.* **142**, 2593. DOI:10.1121/1.5014494.

Ferber, C. and Cabanac, M. (1987). Influence of noise on gustatory affective ratings and preference for sweet or salt, *Appetite* **8**, 229–235.

Fiegel, A., Meullenet, J.-F., Harrington, R. J., Humble, R. and Seo, H.-S. (2014). Background music genre can modulate flavor pleasantness and overall impression of food stimuli, *Appetite* **76**, 144–152.

Fink, D. J. (2017). What is a safe noise level for the public?, *Am. J. Public Health* **107**, 44–45.

Forsyth, A. and Cloonan, M. (2008). Alco-pop? The use of popular music in Glasgow pubs, *Pop. Music Soc.* **31**, 57–78.

Fritz, T. H., Brummerloh, B., Urquijo, M., Wegner, K., Reimer, E., Gutekunst, S., Schneider, L., Smallwood, J. and Villringer, A. (2017). Blame it on the bossa nova: transfer of perceived sexiness from music to touch, *J. Exp. Psychol. Gen.* **146**, 1360–1365.

Frolov, Y. P. (1937). *Fish Who Answer the Telephone, and Other Studies in Experimental Biology.* Trans. S. Graham. Kegan Paul, London, UK.

Gau, R. and Noppeney, U. (2016). How prior expectations shape multisensory perception, *NeuroImage* **124**, 876–886.

Geliebter, A., Pantazatos, S. P., McOuatt, H., Puma, L., Gibson, C. D. and Atalayer, D. (2013). Sex-based fMRI differences in obese humans in response to high vs. low energy food cues, *Behav. Brain Res.* **243**, 91–96.

Getz, L. M. and Kubovy, M. (2018). Questioning the automaticity of audiovisual correspondences, *Cognit.* **175**, 101–108.

Graziose, M. M., Koch, P. A., Wolf, R., Gray, H. L., Trent, R. and Contento, I. R. (2019). Cafeteria noise exposure and fruit and vegetable consumption at school lunch: a cross-sectional study of elementary students, *Appetite* **136**, 130–136.

Guéguen, N., Hélène, L. G. and Jacob, C. (2004). Sound level of background music and consumer behavior: an empirical evaluation, *Percept. Mot. Skills* **99**, 34–38.

Guéguen, N., Jacob, C., Hélène, L. G., Morineau, T. and Lourel, M. (2008). Sound level of environmental music and drinking behavior: a field experiment with beer drinkers, *Alcohol Clin. Exp. Res.* **32**, 1795–1798.

Guetta, R. and Loui, P. (2017). When music is salty: the crossmodal associations between sound and taste, *PLoS One* **12**, e0173366. DOI:10.1371/journal.pone.0173366.

Hauck, P. and Hecht, H. (2019). Having a drink with Tchaikovsky: the crossmodal influence of background music on the taste of beverages, *Multisens. Res.* **32**, 1–24.

Hernando, H. (2014). Bacon sizzling, wine glugging and the opening notes of Coronation Street are sounds which relax us most. Daily Mail Online, 31st July. Available online: http://

www.dailymail.co.uk/news/article-2711962/Bacon-sizzling-wine-glugging-opening-notes-Coronation-Street-sounds-relax-most.html?ITO=1490&ns_mchannel=rss&ns_campaign=1490.

Höchenberger, R. and Ohla, K. (2019). A bittersweet symphony: evidence for taste-sound correspondences without effects on taste quality-specific perception, *J. Neurosci. Res.* **97**, 267–275.

Hockey, G. R. J. (1970). Effect of loud noise on attentional selectivity, *Q. J. Exp. Psychol.* **22**, 28–36.

Holt-Hansen, K. (1968). Taste and pitch, *Percept. Mot. Skills* **27**, 59–68.

Holt-Hansen, K. (1976). Extraordinary experiences during cross-modal perception, *Percept. Motor Skills* **43**, 1023–1027.

Houge, B. and Friedrichs, J. (2013). Food opera: a new genre for audio-gustatory expression, in: *Musical Metacreation: Papers From the 2013 AIIDE Workshop*, Boston, MA. https://www.aaai.org/ocs/index.php/AIIDE/AIIDE13/paper/viewFile/7449/7676.

Hutchison, K. A. (2003). Is semantic priming due to association strength or feature overlap? A microanalytic review, *Psychon. Bull. Rev.* **10**, 785–813.

Kaiser, D., Silberger, S., Hilzendegen, C. and Stroebele-Benschop, N. (2016). The influence of music type and transmission mode on food intake and meal duration: An experimental study, *Psychol. Music* **44**, 1419–1430.

Kantono, K., Hamid, N., Shepherd, D., Yoo, M. J. Y., Grazioli, G. and Carr, B. T. (2016a). Listening to music can influence hedonic and sensory perceptions of gelati, *Appetite* **100**, 244–255.

Kantono, K., Hamid, N., Shepherd, D., Lin, Y. H. T., Yakuncheva, S., Yoo, M. J., Grazioli, G. and Carr, B. T. (2016b). The influence of auditory and visual stimuli on the pleasantness of chocolate gelati, *Food Qual. Prefer.* **53**, 9–18.

Kantono, K., Hamid, N., Shepherd, D., Yoo, M. J. Y., Carr, B. T. and Grazioli, G. (2016c). The effect of background music on food pleasantness ratings, *Psychol. Music* **44**, 1111–1125.

Kantono, K., Hamid, N., Shepherd, D., Lin, Y. H. T., Brard, C., Grazioli, G. and Carr, B. T. (2018). The effect of music on gelato perception in different eating contexts, *Food Res. Int.* **113**, 43–56.

Kantono, K., Hamid, N., Shepherd, D., Lin, Y. H. T., Skiredj, S. and Carr, B. T. (2019). Emotional and electrophysiological measures correlate to flavour perception in the presence of music, *Physiol. Behav.* **199**, 154–164.

Keller, S. and Spence, C. (2017). Sounds delicious: a crossmodal perspective on restaurant atmospherics and acoustical design, *J. Acoust. Soc. Am.* **142**, 2594. DOI:10.1121/1.5014501.

Klapetek, A., Ngo, M. K. and Spence, C. (2012). Does crossmodal correspondence modulate the facilitatory effect of auditory cues on visual search?, *Atten. Percept. Psychophys.* **74**, 1154–1167.

Knöferle, K. M. and Spence, C. (2012). Crossmodal correspondences between sounds and tastes, *Psychon. Bull. Rev.* **19**, 992–1006.

Knoeferle, K. and Spence, C. (in press). Sound in the context of (multi-)sensory marketing, in: *Oxford Handbook of Music and Advertising*, S.-L. Tan (Ed.), Oxford University Press, Oxford, UK.

Knoeferle, K. M., Woods, A., Käppler, F. and Spence, C. (2015). That sounds sweet: using crossmodal correspondences to communicate gustatory attributes, *Psychol. Market.* **32**, 107–120.

Knoeferle, K. M., Knoeferle, P., Velasco, C. and Spence, C. (2016). Multisensory brand search: how the meaning of sounds guides consumers' visual attention, *J. Exp. Psychol. Appl.* **22**, 196–210.

Knoeferle, K. M., Paus, V. C. and Vossen, A. (2017). An upbeat crowd: fast in-store music alleviates negative effects of high social density on customers' spending, *J. Retail.* **93**, 541–549.

Konečni, V. J. (2008). Does music induce emotion? A theoretical and methodological analysis, *Psychol. Aesthet. Creat. Arts* **2**, 115–129.

Kontukoski, M., Luomala, H., Mesz, B., Sigman, M., Trevisan, M., Rotola-Pukkila, M. and Hopia, A. I. (2015). Sweet and sour: music and taste associations, *Nutr. Food Sci.* **45**, 357–376.

Kou, S., McClelland, A. and Furnham, A. (2018). The effect of background music and noise on the cognitive test performance of Chinese introverts and extraverts, *Psychol. Music* **46**, 125–135.

Kupfermann, I. (1964). Eating behaviour induced by sounds, *Nature* **201**, 324.

Labroo, A. A., Dhar, R. and Schwartz, N. (2008). Of frog wines and frowning watches: semantic priming, perceptual fluency, and brand evaluation, *J Consum. Res.* **34**, 819–831.

Lammers, H. B. (2003). An oceanside field experiment on background music effects on the restaurant tab, *Percept. Mot. Skills* **96**, 1025–1026.

Lanza, J. (2004). *Elevator Music: A Surreal History of Muzak, Easy-Listening, and Other Moodsong*. University of Michigan Press, Ann Arbor, MI.

Liew, K., Lindborg, P., Rodrigues, R. and Styles, S. J. (2018). Cross-modal perception of noise-in-music: audiences generate spiky shapes in response to auditory roughness in a novel electroacoustic concert setting, *Front. Psychol.* **9**, 178. DOI:10.3389/fpsyg.2018.00178.

Lindborg, P. (2016). A taxonomy of sound sources in restaurants, *Appl. Acoust.* **110**, 297–310.

Liu, S., Meng, Q. and Kang, J. (2018). Effects of children characteristics on sound environment in fast food restaurants in China, in: *Euronoise 2018*, Crete, Greece.

Lowe, M. L. and Haws, K. L. (2017). Sounds big: the effects of acoustic pitch on product perceptions, *J Mark. Res.* **54**, 331–346.

Lowe, M., Ringler, C. and Haws, K. (2018). An overture to overeating: the cross-modal effects of acoustic pitch on food preferences and serving behavior, *Appetite* **123**, 128–134.

Lucas, M. (2000). Semantic priming without association: a meta-analytic review, *Psychon. Bull. Rev.* **7**, 618–630.

Magnini, V. P. and Thelen, S. T. (2008). The influence of music on perceptions of brand personality, décor, and service quality: The case of classical music in a fine-dining restaurant, *Journal of Hospitality & Leisure Marketing* **16**, 286–300.

Mamalaki, E., Zachari, K., Karfopoulou, E., Zervas, E. and Yannakoulia, M. (2017). Presence of music while eating: effects on energy intake, eating rate and appetite sensations, *Physiol. Behav.* **168**, 31–33.

Marin, M. M., Schober, R., Gingras, B. and Leder, H. (2017). Misattribution of musical arousal increases sexual attraction towards opposite-sex faces in females, *PLoS One* **12**, e0183531. DOI:10.1371/journal.pone.0183531.

Martens, M., Skaret, J. and Lea, P. (2010). Sensory perception of food products affected by different music genres, in: *EuroSense*, Vitoria-Gasteiz, Spain, pp. 1.124.

McCarron, A. and Tierney, K. J. (1989). The effect of auditory stimulation on the consumption of soft drinks, *Appetite* **13**, 155–159.

McClure, S. M., Li, J., Tomlin, D., Cypert, K. S., Montague, L. M. and Montague, P. R. (2004). Neural correlates of behavioral preference for culturally familiar drinks, *Neuron* **44**, 379–387.

McFadden, D., Barr, A. E. and Young, R. E. (1971). Audio analgesia: lack of a cross-masking effect on taste, *Percept. Psychophys.* **10**, 175–179.

Mesz, B., Trevisan, M. A. and Sigman, M. (2011). The taste of music, *Perception* **40**, 209–219.

Mesz, B., Sigman, M. and Trevisan, M. A. (2012). A composition algorithm based on cross-modal taste-music correspondences, *Front. Hum. Neurosci.* **6**, 71. DOI:10.3389/fnhum.2012.00071.

Milliman, R. E. (1986). The influence of background music on the behavior of restaurant patrons, *J. Consumer Res.* **13**, 286–289.

Moir, J. (2015). Why are restaurants so noisy? Can't hear a word your other half says when you dine out? Our test proves restaurants can be a loud as rock concerts, *Daily Mail Online*, December 5th. http://www.dailymail.co.uk/news/article-3346929/Why-restaurants-noisy-t-hear-word-half-says-dine-test-proves-restaurants-loud-rock-concerts.html.

Motoki, K., Saito, T., Nouchi, R., Kawashima, R. and Sugiura, M. (2018). Tasting voices: cross-modal correspondences between tastes and voice pitch increase advertising effectiveness, poster presented at the *Tohoku Forum for Creativity (Acoustical Communication)*, October 21st, Sendai, Japan.

Muniz, R., Harrington, R. J., Ogbeidea, G.-C. and Seo, H.-S. (2017). The role of sound congruency on ethnic menu item selection and price expectations, *Int. J. Hospitality & Tourism Admin.* **18**, 245–271.

Nahorna, O., Berthommier, F. and Schwartz, J.-L. (2012). Binding and unbinding the auditory and visual streams in the McGurk effect, *J. Acoust. Soc. Am.* **132**, 1061. DOI:10.1121/1.4728187.

Nahorna, O., Berthommier, F. and Schwartz, J.-L. (2015). Audio-visual speech scene analysis: characterization of the dynamics of unbinding and rebinding the McGurk effect, *J. Acoust. Soc. Am.* **137**, 362. DOI:10.1121/1.4904536.

Noel, C. and Dando, R. (2015). The effect of emotional state on taste perception, *Appetite* **95**, 89–95.

North, A. C. (2012). The effect of background music on the taste of wine, *Br. J. Psychol.* **103**, 293–301.

North, A. C. and Hargreaves, D. J. (1997). Liking, arousal potential, and the emotions expressed by music, *Scand. J. Psychol.* **38**, 45–53.

North, A. C. and Hargreaves, D. J. (1998). The effects of music on atmosphere and purchase intentions in a cafeteria, *J. Appl. Soc. Psychol.* **28**, 2254–2273.

North, A. C., Hargreaves, D. J. and McKendrick, J. (1997). In-store music affects product choice, *Nature* **390**, 132.

North, A. C., Hargreaves, D. J. and McKendrick, J. (1999). The influence of in-store music on wine selections, *J. Appl. Psychol.* **84**, 271–276.

North, A. C., Shilcock, A. and Hargreaves, D. J. (2003). The effect of musical style on restaurant customers' spending, *Environ. Behav.* **35**, 712–718.

North, A. C., Sheridan, L. P. and Areni, C. S. (2016). Music congruity effects on product memory, perception, and choice, *J. Retail.* **92**, 83–95.

Novak, C. C., La Lopa, J. and Novak, E. R. (2010). Effects of sound pressure levels and sensitivity to noise on mood and behavioral intent in a controlled fine dining restaurant environment, *J. Culinary Sci. & Technol.* **8**, 191–218.

Ofli, F., Aytar, Y., Weber, I., al Hammouri, R. and Torralba, A. (2017). Is Saki# delicious?: the food perception gap on Instagram and its relation to health, in: *Proceedings of the 26th International Conference on World Wide Web*, pp. 509–518.

Padulo, C., Tommasi, L. and Brancucci, A. (2018). Implicit association effects between sound and food images, *Multisens. Res.* **31**, 779–791.

Palmer, S. E., Schloss, K. B., Xu, Z. and Prado-León, L. R. (2013). Music–color associations are mediated by emotion, *Proc. Natl Acad. Sci. U.S.A.* **110**, 8836–8841.

Parise, C. V. (2016). Crossmodal correspondences: standing issues and experimental guidelines, *Multisens. Res.* **29**, 7–28.

Parise, C. and Spence, C. (2013). Audiovisual cross-modal correspondences in the general population, in: *The Oxford Handbook of Synaesthesia*, J. Simner and E. M. Hubbard (Eds), pp. 790–815. Oxford University Press, Oxford, UK.

Pavlov, P. I. (1927). *Conditioned Reflexes: an Investigation of the Physiological Activity of the Cerebral Cortex*. Translated and edited by G. V. Anrep. Oxford University Press, London, UK.

Peeples, L. (2010). Making scents of sounds: noises may alter how we perceive odors, *Sci. Am.* **302**, 28–29.

Pellegrino, R., Luckett, C. R., Shinn, S. E., Mayfield, S., Gude, K., Rhea, A. and Seo, H.-S. (2015). Effects of background sound on consumers' sensory discriminatory ability among foods, *Food Qual. Prefer.* **43**, 71–78.

Petitt, L. A. (1958). The influence of test location and accompanying sound in flavor preference testing of tomato juice, *Food Technol.* **12**, 55–57.

Peynaud, E. (1987). *The Taste of Wine: the Art and Science of Wine Appreciation*. Trans. M. Schuster. Macdonald and Co, London, UK.

Piesse, C. H. (1891). *Piesse's Art of Perfumery*, 5th edn. Piesse and Lubin, London, UK.

Piqueras-Fiszman, B. and Spence, C. (2015). Sensory expectations based on product-extrinsic food cues: an interdisciplinary review of the empirical evidence and theoretical accounts, *Food Qual. Prefer* **40**, 165–179.

Plailly, J., Howard, J. D., Gitelman, D. R. and Gottfried, J. A. (2008). Attention to odor modulates thalamocortical connectivity in the human brain, *J. Neurosci.* **28**, 5257–5267.

Powers, A. R., Mathys, C. and Corlett, P. R. (2017). Pavlovian conditioning-induced hallucinations result from overweighting of perceptual priors, *Science* **357**, 596–600.

Pyko, A., Eriksson, C., Lind, T., Mitkovskaya, N., Wallas, A., Ögren, M., Östenson, C.-G. and Pershagen, G. (2017). Long-term exposure to transportation noise in relation to development of obesity — a cohort study, *Environ. Health Perspect.* **125**, 117005. DOI:10.1289/EHP1910.

Ragneskog, H., Bråne, G., Karlsson, I. and Kihlgren, M. (1996). Influence of dinner music on food intake and symptoms common in dementia, *Scand. J. Caring Sci.* **10**, 11–17.

Rahne, T., Köppke, R., Nehring, M., Plontke, S. K. and Fischer, H.-G. (2018). Does ambient noise or hypobaric atmosphere influence olfactory and gustatory function?, *PLoS One* **13**, e0190837. DOI:10.1371/journal.pone.0190837.

Reinoso Carvalho, F., Van Ee, R., Rychtarikova, M., Touhafi, A., Steenhaut, K., Persoone, D. and Spence, C. (2015a). Using sound-taste correspondences to enhance the subjective value of tasting experiences, *Front. Psychol. Eat. Behav.* **6**, 1309. DOI:10.3389/fpsyg.2015.01309.

Reinoso Carvalho, F., Van Ee, R., Rychtarikova, M., Touhafi, A., Steenhaut, K., Persoone, D., Spence, C. and Leman, M. (2015b). Does music influence the multisensory tasting experience?, *J. Sens. Sci.* **30**, 404–412.

Reinoso Carvalho, F., Van Ee, R. and Rychtarikova, M. (2016a). Matching soundscapes and music with food types, in: *Proceedings of Euroregio*, Porto, Portugal, pp. 178–186.

Reinoso Carvalho, F., Steenhaut, K., van Ee, R., Touhafi, A. and Velasco, C. (2016b). Sound-enhanced gustatory experiences and technology, in: *Proceedings of the 1st Workshop on Multi-Sensorial Approaches to Human-Food Interaction*, Tokyo, Japan, p. 5.

Reinoso Carvalho, F., Velasco, C., Van Ee, R., Leboeuf, Y. and Spence, C. (2016c). Music influences hedonic and taste ratings in beer, *Front. Psychol. Eating Behav.* **7**, 636. DOI:10. 3389/fpsyg.2016.00636.

Reinoso Carvalho, F., Wang, Q. (J.), De Causmaecker, B., Steenhaut, K., Van Ee, R. and Spence, C. (2016d). Tune that beer! Listening to the pitch of beer, *Beverages* **2**, 31. DOI:10.3390/ beverages2040031.

Reinoso Carvalho, F., Wang, Q. (J.), Van Ee, R. and Spence, C. (2016e). The influence of sound-scapes on the perception and evaluation of beers, *Food Qual. Prefer.* **52**, 32–41.

Reinoso Carvalho, F., Wang, Q. (J.), Van Ee, R., Persoone, D. and Spence, C. (2017). "Smooth operator": music modulates the perceived creaminess, sweetness, and bitterness of chocolate, *Appetite* **108**, 383–390.

Reinoso Carvalho, F., Dakduk, S., Wagemans, J. and Spence, C. (2019). Not just another pint! Measuring the influence of the emotion induced by music on the consumer's tasting experience, *Multisens. Res.* DOI:10.1163/22134808-20191374.

Remington, B., Roberts, P. and Glautier, S. (1997). The effect of drink familiarity on tolerance to alcohol, *Addictive Behav.* **22**, 45–53.

Roballey, T. C., McGreevy, C., Rongo, R. R., Schwantes, M. L., Steger, P. J., Wininger, M. A. and Gardner, E. B. (1985). The effect of music on eating behaviour, *Bull. Psychon. Soc.* **23**, 221–222.

Rozin, P. and Rozin, A. (2018). Advancing understanding of the aesthetics of temporal sequences by combining some principles and practices in music and cuisine with psychology, *Perspect. Psychol. Sci.* **13**, 598–617.

Rudmin, F. and Cappelli, M. (1983). Tone-taste synesthesia: a replication, *Percept. Mot. Skills* **56**, 118. DOI:10.2466/pms.1983.56.1.118.

Schietecat, A. C., Lakens, D., IJsselsteijn, W. A. and de Kort, Y. A. W. (2018). Predicting context-dependent cross-modal associations with dimension-specific polarity attributions part 1 — brightness and aggression, *Collabra: Psychol.* **4**, 14. DOI:10.1525/collabra.110.

Seo, H.-S. and Hummel, T. (2011). Auditory-olfactory integration: congruent or pleasant sounds amplify odor pleasantness, *Chem. Senses* **36**, 301–309.

Seo, H.-S. and Hummel, T. (2015). Influence of auditory cues on chemosensory perception, in: *The Chemical Sensory Informatics of Food: Measurement, Analysis, Integration*, B. Guthrie, J. D. Beauchamp, A. Buettner and B. K. Lavine (Eds), *ACS Symposium Series, American Chemical Society*, pp. 41–56. Oxford University Press, Oxford, UK.

Seo, H.-S., Gudziol, V., Hähner, A. and Hummel, T. (2011). Background sound modulates the performance of odor discrimination task, *Exp. Brain Res.* **212**, 305–314.

Seo, H.-S., Hähner, A., Gudziol, V., Scheibe, M. and Hummel, T. (2012). Influence of background noise on the performance in the odor sensitivity task: effects of noise type and extraversion, *Exp. Brain Res.* **222**, 89–97.

Seo, H.-S., Lohse, F., Luckett, C. R. and Hummel, T. (2014). Congruent sound can modulate odor pleasantness, *Chem. Senses* **39**, 215–228.

Septianto, F. (2016). "Chopin" effect? An exploratory study on how musical tempo influence consumer choice of drink with different temperatures, *Asia Pacific Journal of Marketing and Logistics* **28**, 765–779.

Sester, C., Deroy, O., Sutan, A., Galia, F., Desmarchelier, J.-F., Valentin, D. and Dacremont, C. (2013). "Having a drink in a bar": an immersive approach to explore the effects of context on drink choice, *Food Qual. Prefer.* **28**, 23–31.

Silva, D. (2018). *Impact of simultaneous audition of 4 different musical styles on acceptance of snack crackers*. Unpublished manuscript.

Simner, J., Cuskley, C. and Kirby, S. (2010). What sound does that taste? Cross-modal mapping across gustation and audition, *Perception* **39**, 553–569.

Soto-Faraco, S., Lyons, J., Gazzaniga, M., Spence, C. and Kingstone, A. (2002). The ventriloquist in motion: illusory capture of dynamic information across sensory modalities, *Cogn. Brain Res.* **14**, 139–146.

Spence, C. (2011). Crossmodal correspondences: a tutorial review, *Atten. Percept. Psychophys.* **73**, 971–995.

Spence, C. (2012). Auditory contributions to flavour perception and feeding behaviour, *Physiol. Behav.* **107**, 505–515.

Spence, C. (2014a). Noise and its impact on the perception of food and drink, *Flavour* **3**, 9. DOI:10.1186/2044-7248-3-9.

Spence, C. (2014b). Orienting attention: a crossmodal perspective, in: *The Oxford Handbook of Attention*, A. C. Nobre and S. Kastner (Eds), pp. 446–471. Oxford University Press, Oxford, UK.

Spence, C. (2015a). Eating with our ears: assessing the importance of the sounds of consumption to our perception and enjoyment of multisensory flavour experiences, *Flavour* **4**, 3. DOI:10.1186/2044-7248-4-3.

Spence, C. (2015b). Music from the kitchen, *Flavour* **4**, 25. DOI:10.1186/s13411-015-0035-z.

Spence, C. (2016a). Oral referral: on the mislocalization of odours to the mouth, *Food Qual. Prefer.* **50**, 117–128.

Spence, C. (2016b). The neuroscience of flavour, in: *Multisensory Flavor Perception: from Fundamental Neuroscience Through to the Marketplace*, B. Piqueras-Fiszman and C. Spence (Eds), pp. 235–248. Woodhead Publishing, Sawston, UK.

Spence, C. (2017a). Tasting in the air: a review, *Int. J. Gastron. Food Sci.* **9**, 10–15.

Spence, C. (2017b). *Gastrophysics: the New Science of Eating*. Viking Penguin, London, UK.

Spence, C. (2019). Multisensory experiential wine marketing, *Food Qual. Prefer.* **71**, 106–116.

Spence, C. and Deroy, O. (2013a). On why music changes what (we think) we taste, *i-Perception* **4**, 137–140.

Spence, C. and Deroy, O. (2013b). How automatic are crossmodal correspondences?, *Conscious. Cogn.* **22**, 245–260.

Spence, C. and Driver, J. (1997). On measuring selective attention to a specific sensory modality, *Percept. Psychophys.* **59**, 389–403.

Spence, C. and Gallace, A. (2011). Multisensory design: reaching out to touch the consumer, *Psychol. Market.* **28**, 267–308.

Spence, C. and Piqueras-Fiszman, B. (2014). *The Perfect Meal: the Multisensory Science of Food and Dining*. Wiley-Blackwell, Oxford, UK.

Spence, C. and Squire, S. B. (2003). Multisensory integration: maintaining the perception of synchrony, *Curr. Biol.* **13**, R519–R521.

Spence, C. and Wang, Q. (J.) (2015a). Sensory expectations elicited by the sounds of opening the packaging and pouring a beverage, *Flavour* **4**, 35. DOI:10.1186/s13411-015-0044-y.

Spence, C. and Wang, Q. (J.) (2015b). Wine and music (II): can you taste the music? Modulating the experience of wine through music and sound, *Flavour* **4**, 33. DOI:10.1186/s13411-015-0043-z.

Spence, C. and Wang, Q. (J.) (2015c). Wine and music (III): so what if music influences taste?, *Flavour* **4**, 36. DOI:10.1186/s13411-015-0046-9.

Spence, C. and Wang, Q. (J.) (2015d). Wine and music (I): on the crossmodal matching of wine and music, *Flavour* **4**, 34. DOI:10.1186/s13411-015-0045-x.

Spence, C. and Wang, (Q.) J. (2017). Assessing the impact of closure type on wine ratings and mood, *Beverages* **3**, 52. DOI:10.3390/beverages3040052.

Spence, C. and Wang, Q. J. (2018a). On the meaning(s) of complexity in the chemical senses, *Chem. Senses* **43**, 451–461.

Spence, C. and Wang, Q. J. (2018b). What does the term 'complexity' mean in wine, *Int. J. Gastron. Food Sci.* **14**, 45–54.

Spence, C. and Youssef, J. (2016). Constructing flavour perception: from destruction to creation and back again, *Flavour* **5**, 3. DOI:10.1186/s13411-016-0051-7.

Spence, C., Shankar, M. U. and Blumenthal, H. (2011). 'Sound bites': auditory contributions to the perception and consumption of food and drink, in: *Art and the Senses*, F. Bacci and D. Melcher (Eds), pp. 207–238. Oxford University Press, Oxford, UK.

Spence, C., Richards, L., Kjellin, E., Huhnt, A.-M., Daskal, V., Scheybeler, A., Velasco, C. and Deroy, O. (2013). Looking for crossmodal correspondences between classical music and fine wine, *Flavour* **2**, 29. DOI:10.1186/2044-7248-2-29.

Spence, C., Michel, C. and Smith, B. (2014a). Airplane noise and the taste of umami, *Flavour* **3**, 2. DOI:10.1186/2044-7248-3-2.

Spence, C., Puccinelli, N., Grewal, D. and Roggeveen, A. L. (2014b). Store atmospherics: a multisensory perspective, *Psychol. Market.* **31**, 472–488.

Spence, C., Velasco, C. and Knoeferle, K. (2014c). A large sample study on the influence of the multisensory environment on the wine drinking experience, *Flavour* **3**, 8. DOI:10.1186/2044-7248-3-8.

Srinivasan, M. (1955). Has the ear a role in registering flavour?, *Bull. Central Food Technol. Res. Inst. Mysore (India)* **4**, 136.

Stafford, L. D., Fernandes, M. and Agobiani, E. (2012). Effects of noise and distraction on alcohol perception, *Food Qual. Prefer.* **24**, 218–224.

Stafford, L. D., Agobiani, E. and Fernandes, M. (2013). Perception of alcohol strength impaired by low and high volume distraction, *Food Qual. Prefer.* **28**, 470–474.

Suddath, C. (2013). How Chipotle's DJ, Chris Golub, creates his playlists, *Business-week*, October 17th. http://www.businessweek.com/articles/2013-10-17/chipotles-music-playlists-created-by-chris-golub-of-studio-orca.

Taylor, R. (2017). Big Mac or Brahms, sir? McDonald's is pumping out classical music to calm rowdy customers in 24 hr restaurants, *Daily Mail Online*, July 11th. http://www.dailymail.co.uk/news/article-4685726/McDonald-s-pump-classical-music-calm-late-night-diners.html.

Trautmann, J., Meier-Dinkel, L., Gertheiss, J. and Mörlein, D. (2017). Noise and accustomation: a pilot study of trained assessors' olfactory performance, *PLoS One* **12**, e0174697. DOI:10.1371/journal.pone.0174697.

Velasco, C., Balboa, D., Marmolejo-Ramos, F. and Spence, C. (2014). Crossmodal effect of music and odor pleasantness on olfactory quality perception, *Front. Psychol.* **5**, 1352. DOI:10.3389/fpsyg.2014.01352.

Velasco, C., Reinoso Carvalho, F., Petit, O. and Nijholt, A. (2016). A multisensory approach for the design of food and drink enhancing sonic systems, in: *Proceedings of the 1st Workshop on Multi-Sensorial Approaches to Human-Food Interaction*, Tokyo, Japan, p. 7.

Victor, A. (2014). Louis Armstrong for starters, Debussy with roast chicken and James Blunt for dessert: British Airways pairs music to meals to make in-flight food taste better, *Daily Mail Online*, October 15th. http://www.dailymail.co.uk/travel/travel_news/article-2792286/british-airways-pairs-music-meals-make-flight-food-taste-better.html.

Wang, Q. J. (2017). *Assessing the mechanisms behind sound-taste correspondences and their impact on multisensory flavour perception and evaluation*. DPhil Thesis, University of Oxford.

Wang, Q. (J.) and Spence, C. (2015a). Assessing the effect of musical congruency on wine tasting in a live performance setting, *i-Perception* **6**, 2041669515593027. DOI:10.1177/2041669515593027.

Wang, Q. (J.) and Spence, C. (2015b). Assessing the influence of the multisensory atmosphere on the taste of vodka, *Beverages* **1**, 204–217.

Wang, Q. (J.) and Spence, C. (2016). 'Striking a sour note': assessing the influence of consonant and dissonant music on taste perception, *Multisens. Res.* **30**, 195–208.

Wang, Q. (J.) and Spence, C. (2017a). Assessing the role of emotional associations in mediating crossmodal correspondences between classical music and wine, *Beverages* **3**, 1. DOI:10.3390/beverages3010001.

Wang, Q. and Spence, C. (2017b). The role of pitch and tempo in sound–temperature crossmodal correspondences, *Multisens. Res.* **30**, 307–320.

Wang, Q. (J.) and Spence, C. (2017c). Assessing the influence of music on wine perception amongst wine professionals, *Food Sci. Nutr.* **6**, 295–301.

Wang, Q. (J.) and Spence, C. (2018). "A sweet smile": the modulatory role of emotion in how extrinsic factors influence taste evaluation, *Cogn. Emot.* **32**, 1052–1061.

Wang, Q. J. and Spence, C. (2019). Sonic packaging: how packaging sounds influence multisensory product evaluation, in: *Multisensory Packaging: Designing New Product Experiences*, C. Velasco and C. Spence (Eds), pp. 103–125. Palgrave MacMillan, Cham, Switzerland.

Wang, Q. (J.), Woods, A. and Spence, C. (2015). "What's your taste in music?" a comparison of the effectiveness of various soundscapes in evoking specific tastes, *i-Perception* **6**, 2041669515622001. DOI:10.1177/2041669515622001.

Wang, Q. J., Wang, S. and Spence, C. (2016). "Turn up the taste": assessing the role of taste intensity and emotion in mediating crossmodal correspondences between basic tastes and pitch, *Chem. Senses* **41**, 345–356.

Wang, Q. J., Mesz, B. and Spence, C. (2017a). Analysing the impact of music on wine perception via TDS and TI, in: *Poster Presented at 12th Pangborn Sensory Science Symposium*, 20–24 August, Providence, RI, USA.

Wang, Q. J., Mesz, B. and Spence, C. (2017b). Assessing the impact of music on basic taste perception using time intensity analysis, in: *MHFI'17 — Proceedings of the 2nd ACM SIGCHI International Workshop on Multisensory Approaches to Human-Food Interaction, Co-Located With ICMI 2017*, Glasgow, UK, pp. 18–22.

Wang, Q. (J.), Keller, S. and Spence, C. (2017c). Sounds spicy: enhancing the evaluation of piquancy by means of a customised crossmodally congruent soundtrack, *Food Qual. Prefer.* **58**, 1–9.

Wang, Q. J., Knoeferle, K. and Spence, C. (2017d). Music to make your mouth water? Assessing the potential influence of sour music on salivation, *Front. Psychol.* **8**, 638. DOI:10.3389/fpsyg.2017.00638.

Wang, Q. J., Mielby, L. A., Thybo, A. K., Bertelsen, A. S., Kidmose, U., Spence, C. and Byrne, D. V. (2019). Sweeter together? Assessing the combined influence of product intrinsic and extrinsic factors on perceived sweetness of fruit beverages, *J Sens. Stud.* **2019**, e12492. DOI:10.1111/joss.12492.

Wang, Q. J., Frank, M., Houge, B., Spence, C. and LaTour, K. A. (in press). The influence of music on the perception of oaked wines — a tasting room case study in the U.S. Finger Lakes Region, *J Wine Res.*

Wansink, B. and Van Ittersum, K. (2012). Fast food restaurant lighting and music can reduce calorie intake and increase satisfaction, *Psychol. Rep. Hum. Resources Market.* **111**, 228–232.

Watson, Q. J. and Gunter, K. L. (2017). Trombones elicit bitter more strongly than do clarinets: a partial replication of three studies of Crisinel and Spence, *Multisens. Res.* **30**, 321–335.

Wesson, D. W. and Wilson, D. A. (2010). Smelling sounds: olfactory–auditory sensory convergence in the olfactory tubercle, *J. Neurosci.* **30**, 3013–3021.

Wesson, D. W. and Wilson, D. A. (2011). Sniffing out the contributions of the olfactory tubercle to the sense of smell: hedonics, sensory integration, and more?, *Neurosci. Biobehav. Rev.* **35**, 655–668.

Wheeler, E. (1938). *Tested Sentences That Sell.* Prentice Hall, New York, NY, USA.

Wilson, S. (2003). The effect of music on perceived atmosphere and purchase intentions in a restaurant, *Psychol. Music* **31**, 93–112.

Wolpert, R. S. (1990). Recognition of melody, harmonic accompaniment, and instrumentation: musicians vs. nonmusicians, *Music Percept.* **8**, 95–105.

Woods, A. T., Poliakoff, E., Lloyd, D. M., Dijksterhuis, G. B. and Thomas, A. (2010). Flavor expectation: the effects of assuming homogeneity on drink perception, *Chemosens. Percept.* **3**, 174–181.

Woods, A. T., Poliakoff, E., Lloyd, D. M., Kuenzel, J., Hodson, R., Gonda, H., Batchelor, J., Dijksterhuis, G. B. and Thomas, A. (2011a). Effect of background noise on food perception, *Food Qual. Prefer.* **22**, 42–47.

Woods, A. T., Lloyd, D. M., Kuenzel, J., Poliakoff, E., Dijksterhuis, G. B. and Thomas, A. (2011b). Expected taste intensity affects response to sweet drinks in primary taste cortex, *Neuroreport* **22**, 365–369.

Yan, K. S. and Dando, R. (2015). A crossmodal role for audition in taste perception, *J. Exp. Psychol. Hum. Percept. Perform.* **41**, 590–596.

Yeoh, J. P. S. and North, A. C. (2010). The effects of musical fit on choice between two competing foods, *Music. Sci.* **14**, 165–180.

Youssef, J., Youssef, L., Juravle, G. and Spence, C. (2017). Plateware and slurping influence regular consumers' sensory discriminative and hedonic responses to a hot soup, *Int. J. Gastron. Food Sci.* **9**, 100–104.

Zampini, M. and Spence, C. (2004). The role of auditory cues in modulating the perceived crispness and staleness of potato chips, *J. Sens. Stud.* **19**, 347–363.

Zellner, D., Geller, T., Lyons, S., Pyper, A. and Riaz, K. (2017). Ethnic congruence of music and food affects food selection but not liking, *Food Qual. Prefer.* **56**(Part A), 126–129.

Ziv, N. (2018). Musical flavor: the effect of background music and presentation order on taste, *Eur. J. Market.* **52**, 1485–1504.

Variations in Food Acceptability with Respect to Pitch, Tempo, and Volume Levels of Background Music

Alexandra Fiegel [1], **Andrew Childress** [2], **Thadeus L. Beekman** [1] and
Han-Seok Seo [1,*]

[1] Department of Food Science, University of Arkansas, 2650 North Young Avenue,
Fayetteville, AR 72704, USA
[2] Department of Music, University of Arkansas, Music Building 201, Fayetteville,
AR 72701, USA

Abstract

This study aimed to determine whether pitch, tempo, and volume levels of music stimuli affect sensory perception and acceptance of foods. A traditional music piece was arranged into versions at two pitches, two tempos, and two volumes. For each session, chocolate and bell peppers were presented for consumption under three sound conditions: 1) upper or 2) lower level with respect to each of the three music elements, and 3) silence. Over three sessions, participants evaluated flavor intensity, pleasantness of flavor, texture impression, and overall impression of food samples, in addition to the pleasantness and stimulation evoked by the music stimuli. Results showed that lower-pitched and louder music stimuli increased hedonic impressions of foods compared to their respective counterparts and/or the silent condition. While the effects of music element levels on hedonic impressions differed with the type of food consumed, the participants liked the foods more when music stimuli were perceived as more pleasant and stimulating. Flavor was perceived as more intense when participants were more stimulated by the music samples. Although a specific element of music stimuli was manipulated, perceptions of other elements also varied, leading to large variations in the music-evoked pleasantness and stimulation. In conclusion, the findings provide empirical evidence that hedonic impressions of foods may be influenced by emotions evoked by music selections varying in music element levels, but it should be also noted that the influences were food-dependent and not pronounced.

Keywords

Music, pitch, tempo, volume, acceptance, flavor, texture

---

* To whom correspondence should be addressed. E-mail: hanseok@uark.edu

1. Introduction

Most people are used to consuming food and beverages in the presence of not only sounds elicited by mastication or swallowing processes, but also external sounds from their surroundings, such as background music or conversations. A recent survey reported that only 3.7% of the 244 U.S. adults preferred eating in silence, while 58.8% of them preferred eating while having conservations with others (Pellegrino *et al.*, 2015). Only 6.2% of respondents reported to prefer eating while listening to music (Pellegrino *et al.*, 2015); however, people are accustomed to eating meals in the presence of background music, most notably at restaurants or foodservice facilities (for a review, Spence, 2014, 2015).

In locations where user consumption is responsible for driving a company's sales and overall stature, atmospherics may play a critical role in shaping a consumer's opinion of a product (Donovan and Rossiter, 1982; Spence *et al.*, 2014). In environments where music contributes to the atmospherics, it is not only important for business owners and marketers to understand the popular trends of music, but it is of greater importance to understand how the background music influences consumers' shopping (or eating) behaviors (Garlin and Owen, 2006; Spence *et al.*, 2014; Yalch and Spangenberg, 1990). The Mehrabian–Russell (MR) model (Mehrabian and Russell, 1974) posits that individuals may respond emotionally to environmental stimuli (e.g., background music or noise), thus leading to approach–avoidance behavior, which potentially helps explain such music-induced behavioral responses (Fiegel *et al.*, 2014). In the MR model, approach behaviors are considered to include four aspects: a desire or willingness to 1) physically stay in the environment, 2) look around and explore the environment, 3) communicate with others in the environment, and 4) increase satisfaction of tasks performed in the environment, with avoidance behaviors displaying opposite effects (Donovan and Rossiter, 1982). In this sense, background music or noise has been found to affect consumers' approach–avoidance behaviors with respect to the amount consumed (Guéguen *et al.*, 2004, 2008; Stroebele and de Castro, 2006), the dollar amount spent (Biswas *et al.*, 2019; Herrington, 1996; Jacob, 2006; Sullivan, 2002), the consumption rate (Guéguen *et al.*, 2008; Milliman, 1986), and the food or beverage items evaluated (Fiegel *et al.*, 2014; Kantono *et al.*, 2016a; Reinoso Carvalho *et al.*, 2016, 2017; Spence, 2015).

Previous research has shown how music genre affects consumption amount (Engels *et al.*, 2012) and menu items selection (Muniz *et al.*, 2017; Zellner *et al.*, 2017). In addition, music genre has been found to affect hedonic ratings of food samples consumed (Fiegel *et al.*, 2014; also see Kantono *et al.*, 2016a, b, c). For example, in our previous study (Fiegel *et al.*, 2014), when participants were asked to rate sensory perception and impression of milk chocolate and bell peppers while listening to different genres of background music: classical, jazz, hip-hop, and rock versions of a traditional music piece ("Air on the G string"), they liked food samples significantly more in the presence of the jazz stimulus than the hip-hop stimulus. In that study, although music-evoked emotions such as pleasantness (valence) and stimulation (arousal) were found to influence overall impression of the food samples, there was no attempt to identify the music elements that played a major role in the music-evoked emotions and the corresponding food acceptability. While genre of ambient music is important in designing appropriate atmospherics, it is not solely responsible for elicited emotional behavior. In other words, the structural elements in a genre might be the driving forces. By considering two points: 1) a cause-and-effect relationship between music elements and behavioral responses (Sullivan, 2002) and 2) emotional influences on sensory perception, pleasantness, and food choice (Gibson, 2006; Noel and Dando, 2015; Pollatos *et al.*, 2007), the present study focuses on the impact of music elements on consumers' food perception and acceptance. The music stimulus ("Air on the G string") used in the present study was modified into upper and lower levels with respect to each element of music: pitch, tempo, and volume. It would be expected that such upper and lower levels might induce contrasting music-evoked emotions and lead to varied approach–avoidance behaviors (Mehrabian and Russell, 1974), resulting in differences with respect to food perception and acceptance.

While there is not a consensus about the principal constituent elements of music (Burton, 2015), 'pitch', 'tempo', and 'volume' were chosen as elements of music for manipulating in this study. The three elements have been found to differ in the music-evoked emotions between the upper and lower levels of each element (Berger and Schneck, 2003; Gomez and Danuser, 2007; Sullivan, 2002; Webster and Weir, 2005). The first music element investigated was the relative high or low frequency of the stimulus, characterized by its pitch (Bruner, 1990). Earlier studies demonstrated that high pitches, wide pitch ranges, and large pitch variations may be related to emotions of high activation, excitement, anger, fear, happiness, and pleasantness, while low pitches, narrow pitch ranges, and small pitch variations may be associated with emotions of low activation, calmness, sadness, unpleasantness, and pleasantness (Balkwill and Thompson, 1999; Gabrielsson and Lindström, 2001; Hevner, 1937; Yang and Kang, 2016). Notably, it has been reported that both high- and low-pitched stimuli are associated with 'pleasantness', indicating a probable dependence on the combination and interaction with other music parameters (Gabrielsson and Lindström, 2001; Yang and Kang, 2016). In addition, differences in pitch level have been found to affect food perception and acceptance. For example, high-pitched musical selections tend to enhance both sweet and sour taste perceptions, while low-pitched musical selections tend to induce decreased bitter and salty perceptions (Crisinel and Spence, 2010; Mesz *et al.*,

2011; for a review, Knöferle and Spence, 2012). High-pitched sound stimuli have also been found to be associated with higher intensities of crispiness (Zampini and Spence, 2004; also see Vickers, 1991) and carbonation (Zampini and Spence, 2005).

Tempo, musical rate progression, has been investigated in a variety of research areas (Milliman, 1986; Webster and Weir, 2005; Wilson, 2003). To accommodate context-induced tempo variations, a range of 70–110 beats per minute (bpm) has been suggested as appropriate (Bruner, 1990), with tempos below and above these values considered slow and fast, respectively. Variations in the tempo of a sound stimulus may result in differences in the music-evoked emotions (Hevner, 1937; Khalfa *et al.*, 2008; Webster and Weir, 2005). Faster tempos of sound stimuli tend to be associated with pleasantness and the emotion of happiness, compared to slower tempos likely being related to the emotions of sadness, tranquility, and sentimentality (Hevner, 1937; Webster and Weir, 2005). Additionally, previous studies have emphasized the tempo effect of the sound stimulus on changes in consumer behaviors related to eating rather than food perception and acceptance (Milliman, 1982, 1986; Stroebele and de Castro, 2006). For example, chewing behavior can be affected by the tempo of a sound stimulus, as shown in a study where individuals increased their number of bites per minute as tempo increased, and decreased that number as tempo decreased (Roballey *et al.*, 1985; see also McElrea and Stranding, 1992).

The amplitude of a sound stimulus (wave) is commonly called its volume (or sound pressure level) (Novak *et al.*, 2010). While high volume of the sound stimulus is likely to be associated with joy, excitement, fear, and anger emotions, low volume of the sound stimulus tends to be associated with calm sad, solitude, peaceful, and spiritual emotions (Berger and Schneck, 2003; also see Bruner, 1990). Volume (loudness) levels of the sound stimulus have been found to alter sensory perception and acceptance of food stimuli (Ferber and Cabanac, 1987; Spence, 2012; Stafford *et al.*, 2012; Woods *et al.*, 2011; Yan and Dando, 2015). For example, Woods *et al.* (2011) showed that loud background noise (75–85 decibels [dB]) could increase crunchiness, while concurrently decreasing sweetness and saltiness perceptions of food samples when compared to quiet background noise (45–55 dB).

Our previous study has shown that the influence of background music on overall impression of food stimuli was more pronounced in 'milk chocolate' than in sliced bell peppers. This result was probably due to milk chocolate being considered an 'emotional food', making it more suitable to mediate music-evoked emotions than bell peppers being considered a 'non-emotional food' (Fiegel *et al.*, 2014). It is worth noting that people are likely to consume energy dense and sweet foods when they want to lessen mental stress. In this way, an influence of music-induced stress (e.g., loud music; Ferber and Cabanac, 1987) on sensory perception and acceptance may differ between the emotional and non-emotional foods. Therefore, it is worth exploring whether the effects of music element levels on sensory perception and acceptance can differ as a function of food type.

As mentioned earlier, the overall objective of the present study was to determine whether upper/lower levels of music elements (in particular pitch, tempo, and volume) could influence sensory perception and acceptance for foods consumed while listening, as well as ascertaining whether such an influence might change as a function of food type (Study 1). The specific element arranged into either upper or lower level may affect subjective perceptions of other elements because elements of music stimulus interact with one another (Gabrielsson and Lindström, 2001; Yang and Kang, 2016). Therefore, Study 2 aimed to determine whether music stimuli arranged into two levels, with respect to a specific element of music, could affect subjective perceptions of pitch, tempo, and volume (loudness), as well as pleasantness and stimulation evoked by the music stimuli (Study 2).

2. Materials and Methods

This study was conducted in accordance with the Declaration of Helsinki for studies on human subjects, and the protocols (#1204663 and #1710076017) were approved by the Institutional Review Board of the University of Arkansas (Fayetteville, AR, USA). The experimental procedure was explained to all participants and a written informed consent was obtained prior to their participation.

2.1. Study 1

2.1.1. Participants
Sixty-one healthy U.S. adults (46 females and 15 males; 58 Caucasians, two African Americans, and one American Indian or Alaskan Native) with ages ranging from 18 to 30 years (mean age $\pm$ standard deviation [SD] $= 26 \pm 3$ years) were recruited through the consumer profile database of the University of Arkansas Sensory Service Center (Fayetteville, AR, USA), which is comprised of more than 5000 Northwest Arkansas residents. Earlier research showed that 40 to 100 panelists is an adequate sample size in consumer testing. This size allows a stability of sample difference and a repeatability of the significance level to be generated in consumer testing (Gacula Jr. and Rutenbeck, 2006; also see Mammasse and Schlich, 2014; Moskowitz, 1997). These publications support that the number of participants in this study ($N = 61$) was adequate. Each participant self-reported having no clinical history of major diseases, such as diabetes, cancer, cardiovascular disease, or renal disease.

In addition, each panelist reported no impairments in his/her senses of smell, taste, and hearing.

2.1.2. Food Samples and Presentation

Selection of food samples used in this study, milk chocolate (The Hershey Co., Hershey, PA, USA) and red bell peppers (Walmart Distributor, Bentonville, AR, USA), was based on our previous study (Fiegel *et al.*, 2014). Chocolate and bell peppers were considered to be emotional and non-emotional foods, respectively (Fiegel *et al.*, 2014).

The food samples were purchased from a local supermarket one day before the study. The chocolate was stored at room temperature (approximately 20°C) and broken into bite-sized pieces (3.5 × 2.0 × 0.5 cm) prior to serving. The bell peppers were stored in a refrigerator (4°C) and taken out one hour prior to evaluation to allow them to reach room temperature, then sliced into bite-sized pieces (3.5 × 3.0 × 0.5 cm).

2.1.3. Music Stimuli and Presentation

As an auditory cue, the traditional music piece: "Air on the G string" (an adaptation by August Wilhelmj of the Air, the second movement in the Orchestral Suite No. 3 in D major, BWV 1068, composed by Johann Sebastian Bach) was edited to highlight the lower and upper levels of three musical elements: pitch (lower versus higher), tempo (slower versus faster), and volume (quieter versus louder). The six music stimuli (three music elements × two levels) were edited using the following sound arrangement software: Ableton Live (Ableton Inc., Berlin, Germany), Squidfont Orchestral (http://soundfonts. darkesword.com), and MacMP3Gain (Berry Rinaldo, http://www.rinaldo.net/ Mac_Archive/AudioTron/MacMP3Gain). Table 1 lists the characteristics of the arranged music stimuli, with each divided into two different levels with respect to each of the three music elements: pitch, tempo, and volume. For the level of volume, a slower music stimulus was presented at either quiet or loud level (Table 1). Using a sound editor program (Power Sound Editor Free, ver. 6.9.6, PowerSE Co., Ltd.), each of the six music stimuli was edited to an interval of approximately three minutes and administered through headphones. Music stimuli are available upon request directed to the corresponding author.

2.1.4. Procedure

This study was conducted over a span of three days (i.e., three test sessions) and all participants attended at the same time on all three days. During each test session, one of the three music element conditions (i.e., pitch, tempo, or volume) was randomly presented to participants. More specifically, during each test session, all participants received six pairings of food and music stimuli (i.e., two food samples [chocolate and bell peppers] by three music stimuli [silence, lower level, and upper level] of each of the three music elements).

Table 1.
Element characteristics of music stimuli used in this study

Pitch		Tempo		Volume[1]	
Lower	Higher	Slower	Faster	Quieter	Louder
A2 (110.0 Hz)	A5 (880.0 Hz)	40 bpm	92 bpm	47.0 dB	61.2 dB
to	to			to	to
C4 (261.6 Hz)	C7 (2093.0 Hz)			66.9 dB	90.2 dB

[1]Volume range of music stimulus was measured from one side of the headphones (Model MDR-V150, Sony, Tokyo, Japan), using a sound level meter (R8080, REED Instruments, Wilmington, NC, USA).

These pairings were presented in a sequential monadic fashion as specified by the Williams Latin Square Design (Williams, 1949).

Participants were seated in individual sensory booths on chairs located 70 cm from a computer monitor. They were asked to place headphones (Model MDR-W08L, Sony, Tokyo, Japan) snugly on their ears until completion of the entire test during each test session. Instructions and scales were presented using a computerized sensory data acquisition system, Compusense® five (Release 4.6-SP3, Compusense Inc., Guelph, ON, Canada). Participants were asked to taste either milk chocolate or bell peppers placed in a soufflé cup labelled with a three-digit identification code either in the presence of a music stimulus or in a silent condition. For each music stimulus condition, participants received one type of food sample 30 s after initial music presentation and they consumed it in the presence of the music stimulus. In a silent condition, participants consumed a food sample in the absence of music stimulus.

After consuming a food sample, each participant rated flavor intensity and pleasantness on two 15-cm line scales ranging from 0 (extremely weak; extremely unpleasant) to 15 (extremely strong; extremely pleasant). They also rated each food sample with respect to texture impression and overall impression on two 15-cm line scales ranging from 0 (extremely dislike) to 15 (extremely like). Finally, participants rated the pleasantness (also referred to as valence) and stimulation (also referred to arousal or activation) of the music stimuli presented on two vertical nine-point category scales ranging from 1 (extremely unpleasant; extremely non-stimulating) to 9 (extremely pleasant; extremely stimulating), which were used in our previous study (Fiegel *et al.*, 2014). A non-verbal pictorial assessment technique such as the Self-Assessment Manikin scale (Bradley and Lang, 1994) has also been used in other studies to assess participants' affective responses, such as valence, arousal, and dominance, to the music stimuli (Kantono *et al.*, 2016a; Septianto, 2016). The time interval between sample presentations was 60 s. During each break, spring water (Clear Mountain Spring Water, Taylor Distributing, Heber Springs, AR, USA) for palate cleansing was presented in silence.

2.1.5. Data Analysis

Data were analyzed using SPSS 25.0 for Windows™ (IBM SPSS Inc., Chicago, IL, USA). Since the Shapiro–Wilk Normality test (Shapiro and Wilk, 1965) revealed that the null hypothesis, i.e., that data came from a normally distributed population, was rejected ($P < 0.05$), non-parametric statistical methods were applied for subsequent data analysis.

To determine influences of music element levels with respect to pitch, tempo, or volume on flavor intensity, flavor pleasantness, texture impression, or overall impression of chocolate or bell peppers, a non-parametric Friedman test (Friedman, 1937) was used. To measure an effect size for the Friedman test, Kendall's coefficient (W) of concordance (Kendall and Smith, 1939) was used (Tomczak and Tomczak, 2014). Kendall's coefficient of concordance ranges from 0 (indicating no agreement between evaluators with respect to ranked data) to 1 (indicating a perfect agreement) (Tomczak and Tomczak, 2014). When the Friedman test indicated a statistical difference among the three sound conditions of each music element, the Wilcoxon signed-rank test was used for individual paired comparisons.

As described above, a total of six music stimuli varying in pitch, tempo, and volume were presented in this study. Six stimuli were compared with respect to their effects on sensory perception and impression of food samples, aiming to identify the element level with the greatest impact. Because all six music stimuli were presented over a span of three days, rating differences between the music stimuli and silent condition with respect to sensory perception or impression were compared among the six music stimuli conditions. More specifically, for each element of music, intensity or impression ratings of food samples under the silent condition were subtracted from those ratings in the presence of the music stimulus (e.g., flavor intensity rating of chocolate in the presence of higher-pitched music minus those of chocolate under the silent condition). Those subtracted values were then compared using the Friedman test. It should be noted that during each test session, participants evaluated both food samples in a silent condition.

The Wilcoxon signed-rank test was also used to determine whether ratings of pleasantness or stimulation for music stimuli might differ with respect to lower and upper levels of each music element. A correlation analysis was used to determine whether individual ratings of music pleasantness (or stimulation) could be associated with ratings of flavor intensity, flavor pleasantness, texture impression, or overall impression of food samples. Since there was a correlation between individual ratings of music pleasantness and stimulation,

Spearman partial correlation analysis with covariance as ratings of music stimulation (or pleasantness) was used. A statistical significance was determined when $P < 0.05$.

2.2. Study 2

2.2.1. Participants
One hundred twenty-three healthy U.S. adults (65 females and 58 males; 92 Caucasians, 14 Asians, 10 African Americans, 4 Latinos, 2 Middle Easterners, and 1 American Indian or Alaskan Native) with ages ranging from 19 to 73 years [mean age ± standard deviation (SD) = 43 ± 15 years] were recruited through the consumer profile database of the University of Arkansas Sensory Service Center. Each participant self-reported having neither a clinical history of major diseases nor any impairments in his/her senses of smell, taste, and hearing.

2.2.2. Music Stimuli and Presentation
The six music stimuli presented in Study 1 (Table 1) were also used in Study 2. No silent condition was used in this study.

2.2.3. Procedure
During a test session, participants were seated in individual sensory booths on chairs and asked to place headphones (Model MDR-V150, Sony, Tokyo, Japan) snugly on their ears. They were instructed to refrain from removing their headphones until completion of the entire test. Prior to a main study, a brief instruction about both experimental design and how to use scales was given with a warm-up music stimulus (Salut d'Amour in E Major Op. 12, composed by Edward Elgar; performed by Classic Essay, Melon, Seoul, Republic of Korea). More specifically, while listening to the warm-up music stimulus, participants were asked to rate how they perceived the levels of pitch, tempo, and volume on three nine-point category scales: 1 = extremely low-pitched/extremely slow/extremely quiet, 2 = very low-pitched/very slow/very quiet, 3 = moderately low-pitched/moderately slow/moderately quiet, 4 = slightly low-pitched/slightly slow/slightly quiet, 5 = neither low-pitched nor high-pitched/neither slow nor fast/neither quiet nor loud, 6 = slightly high-pitched/slightly fast/slightly loud, 7 = moderately high-pitched/moderately fast/moderately loud, 8 = very high-pitched/very fast/very loud, and 9 = extremely high-pitched/extremely fast/extremely loud. In addition, participants rated the pleasantness (also referred to as valence) and stimulation (also referred to arousal or activation) of the warm-up music stimulus on two vertical nine-point category scales ranging from 1 (extremely unpleasant; extremely non-stimulating) to 9 (extremely pleasant; extremely stimulating).

Following a warm-up music stimulus, all three subsets (i.e., elements) of music stimuli: pitch (consisting of lower-pitched and high-pitched stimuli), tempo (consisting of slower tempo and faster tempo stimuli), and volume (consisting of quieter and louder stimuli) were presented in a random order during a test session. Within each subset (element), the two music stimuli (e.g., for the subset of pitch, lower-pitched and higher-pitched stimuli) were also randomly presented. While each music stimulus was presented for 60 s via headphones, participants were asked to rate the music stimulus in paper-and-pencil form as they had for the warm-up sample. The time interval between music stimulus presentations was 120 s. Please note that no food samples were served in Study 2.

2.2.4. Data Analysis

Data were analyzed using SPSS 25.0 for Windows™ (IBM SPSS Inc., Chicago, IL, USA) and XLSTAT (Addinsoft, 2019). Since the Shapiro–Wilk Normality test (Shapiro and Wilk, 1965) revealed that the null hypothesis, i.e., that data came from a normally distributed population, was rejected ($P < 0.05$), non-parametric statistical methods were applied for subsequent data analysis.

For each individual element of music stimulus, the Wilcoxon signed-rank test was used to determine whether participants perceived noticeable differences between the two element levels. The two music stimuli, within each element, were also compared with respect to pleasantness and stimulation of music. A statistical significance was determined when $P < 0.05$.

3. Results

3.1. Study 1

3.1.1. Effects of Musical Pitch Levels on Sensory Perception and Impression of Food Samples

3.1.1.1. *Chocolate.*   As shown in Fig. 1A, while the effect size (Kendall's W coefficient) was minimal, there was a significant difference between the texture impression ratings among the three sound conditions ($\chi^2 = 6.10$, $P = 0.04$, $W = 0.05$). More specifically, chocolate texture was liked significantly more in the presence of lower-pitched music than in the presence of higher-pitched music ($P = 0.01$). No significant effects caused by musical pitch levels were found for chocolate samples with respect to flavor intensity ($\chi^2 = 0.34$, $P = 0.84$), flavor pleasantness ($\chi^2 = 3.91$, $P = 0.14$), or overall impression ($\chi^2 = 0.95$, $P = 0.62$) (Fig. 1A).

3.1.1.2. *Bell Peppers.*   As shown in Fig. 1B, while the effect size was minimal, the overall impression of bell peppers was found to differ significantly among three sound conditions ($\chi^2 = 6.12$, $P = 0.047$, $W = 0.05$), indicating that participants overall liked bell peppers significantly more in the presence of lower-pitched music than under a silent condition ($P = 0.02$). However, there

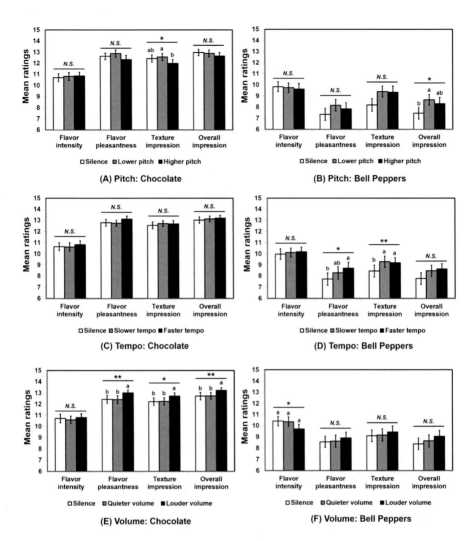

Figure 1. Mean ratings of flavor intensity, flavor pleasantness, texture impression, and overall impression for chocolate (A, C, and E) and bell peppers (B, D, and F) among the three sound conditions: (1) silence, (2) lower or (3) upper level with respect to each of the three music elements (pitch, tempo, and volume) (Study 1). *N.S.* indicates no significance at $P < 0.05$. * and ** indicate significance at $P < 0.05$ and $P < 0.01$, respectively. When the Friedman test indicated a significant difference among the three sound conditions, the Wilcoxon signed-rank test was used for further analysis of individual paired comparisons. Mean ratings with different letters within each category represent a significant difference at $P < 0.05$. While post-hoc tests revealed that flavor intensity ratings of bell peppers (F) were marginally lower in the presence of louder music than in the presence of quieter music ($P = 0.05$) or under a silent condition ($P = 0.06$), such trends were not statistically significant at $P < 0.05$. Error bars represent the standard errors of the means.

was no significant difference between lower-pitched and higher-pitched music stimuli with respect to overall impression of bell peppers ($P = 0.37$). As shown in Fig. 1B, there were no significant effects of musical pitch on flavor intensity ($\chi^2 = 0.92$, $P = 0.63$), flavor pleasantness ($\chi^2 = 0.98$, $P = 0.61$), and texture impression ($\chi^2 = 0.93$, $P = 0.63$) of bell peppers.

3.1.2. Effects of Musical-Tempo Levels on Sensory Perception and Impression of Food Samples

3.1.2.1. *Chocolate.*   As shown in Fig. 1C, there were no significant effects of musical tempo on flavor intensity ($\chi^2 = 0.14$, $P = 0.93$), flavor pleasantness ($\chi^2 = 0.87$, $P = 0.65$), texture impression ($\chi^2 = 1.08$, $P = 0.58$), and overall impression ($\chi^2 = 0.13$, $P = 0.94$) of chocolate sample.

3.1.2.2. *Bell Peppers.*   As shown in Fig. 1D, while the effect sizes were small, flavor pleasantness ($\chi^2 = 8.44$, $P = 0.02$, $W = 0.07$) and texture impression ($\chi^2 = 9.56$, $P = 0.008$, $W = 0.08$) of bell peppers differed significantly among the three sound conditions. More specifically, participants rated bell pepper flavor more pleasant in the presence of faster tempo music than in a silent condition ($P = 0.003$). In addition, participants liked the texture of bell peppers significantly more in the presence of slower tempo ($P = 0.013$) or faster tempo ($P = 0.028$) music than in a silent condition. However, flavor intensity ($\chi^2 = 1.04$, $P = 0.59$) and overall impression ($\chi^2 = 4.83$, $P = 0.09$) of bell peppers did not differ significantly under the three sound conditions.

3.1.3. Effects of Music Volume Levels on Sensory Perception and Impression of Food Samples

3.1.3.1. *Chocolate.*   Figure 1E shows that flavor pleasantness ($\chi^2 = 11.21$, $P = 0.004$, $W = 0.09$), texture impression ($\chi^2 = 6.37$, $P = 0.04$, $W = 0.05$), and overall impression ($\chi^2 = 10.82$, $P = 0.004$, $W = 0.09$) of chocolate sample differed significantly among the three sound conditions. In the presence of louder music, flavor pleasantness, texture impression, and overall impression of chocolate the sample were rated significantly higher than in the presence of quieter music (flavor pleasantness: $P = 0.008$; texture impression: $P = 0.007$; and overall impression: $P = 0.006$) or under the silent condition (flavor pleasantness: $P = 0.007$; texture impression: $P = 0.03$; and overall impression: $P = 0.006$). However, there was no significant effect of music volume level on flavor intensity of chocolate sample ($\chi^2 = 2.87$, $P = 0.24$).

3.1.3.2. *Bell Peppers.*   Figure 1F shows that flavor intensity of bell peppers differed significantly among the three sound conditions ($\chi^2 = 8.69$, $P = 0.01$, $W = 0.07$). While post-hoc tests revealed that flavor intensity ratings of bell peppers were marginally lower in the presence of louder music than in the presence of quieter music ($P = 0.05$) or under a silent condition ($P = 0.06$), such trends were not statistically significant at $P < 0.05$.

In addition, flavor pleasantness ($\chi^2 = 1.47$, $P = 0.48$), texture impression ($\chi^2 = 0.34$, $P = 0.84$), and overall impression ($\chi^2 = 3.02$, $P = 0.22$) of bell peppers did not significantly differ as a function of music volume levels.

3.1.4. Variations in Sensory Perception and Impression of Food Samples as a Function of the Element Level of Music

To identify the music element level that could increase sensory perception and impression of food samples (compared to under a silent condition) the most, the music-induced sensory perception and impression of food samples were compared among all six music stimulus conditions.

3.1.4.1. Chocolate. Figure 2A shows that the music-induced flavor intensity was not significantly different among all six music stimulus conditions ($\chi^2 = 2.48$, $P = 0.78$). Flavor pleasantness of chocolate significantly differed among the six music stimulus conditions ($\chi^2 = 16.47$, $P = 0.006$, $W = 0.05$). Post-hoc tests, as displayed in Fig. 2B, revealed that music-induced flavor pleasantness of chocolate was significantly higher in the presence of the louder music stimulus than in the presence of the higher-pitched music ($P = 0.006$), slower tempo music ($P = 0.006$), or quieter music stimulus ($P = 0.008$). In addition, texture impression of bell peppers significantly differed among the six music stimulus conditions ($\chi^2 = 12.17$, $P = 0.03$, $W = 0.04$). As shown in Fig. 2C, post-hoc tests found that music-induced texture impression of chocolate was significantly higher in the presence of a louder music stimulus than in the presence of higher-pitched music ($P = 0.01$), faster tempo music ($P = 0.03$), or quieter music ($P = 0.007$). In addition, music-induced texture impression of chocolate was significantly higher in the presence of low-pitched music than in that of the high-pitched music ($P = 0.01$). However, music-induced overall impression was not different among the six music stimulus conditions ($\chi^2 = 7.51$, $P = 0.19$) (Fig. 2D).

3.1.4.2. Bell Peppers. Figure 3 shows no significant differences among the six music stimulus conditions with respect to the music-induced flavor intensity ($\chi^2 = 6.33$, $P = 0.28$), flavor pleasantness ($\chi^2 = 4.31$, $P = 0.51$), texture impression ($\chi^2 = 2.99$, $P = 0.70$), and overall impression ($\chi^2 = 1.80$, $P = 0.88$).

3.1.5. Ratings of Pleasantness and Stimulation for Music Stimuli Arranged into Two Levels as a Function of Music Elements: Pitch, Tempo, and Volume

3.1.5.1. Pitch. Wilcoxon signed-rank tests revealed that the pleasantness and stimulation ratings for the pitch-variable music stimuli were both significantly different. The lower-pitched music was rated as significantly more pleasant ($Z = -5.03$, $P < 0.001$) and stimulating ($Z = -2.55$, $P = 0.01$) than the higher-pitched music stimulus (Table 2).

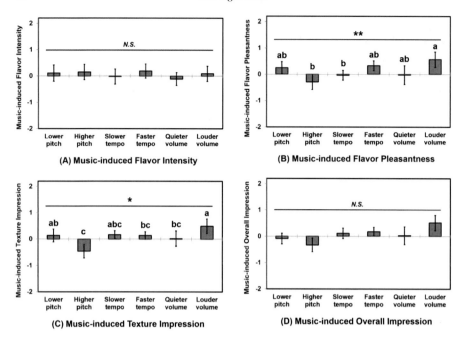

Figure 2. Mean ratings of music-induced flavor intensity (A), music-induced flavor pleasant-ness (B), music-induced texture impression (C), and music-induced overall impression (D) of a chocolate sample with respect to element levels of music (Study 1). For music-induced sen-sory perception or impression, the intensity or impression ratings of the chocolate sample under the silent condition were subtracted from those ratings in the presence of a music stimulus (e.g., flavor intensity rating of chocolate in the presence of higher-pitched music minus those of chocolate under the silent condition). It should be noted that during each test session, par-ticipants evaluated a chocolate sample in a silent condition. Those subtracted values were then compared using the Friedman test. *N.S.* indicates no significance at $P < 0.05$. * and ** indicate significance at $P < 0.05$ and $P < 0.01$, respectively. When the Friedman test indicated a sig-nificant difference among the three sound conditions, the Wilcoxon signed-rank test was used for further analysis of individual paired comparisons. Mean ratings with different letters within each category represent a significant difference at $P < 0.05$. Error bars represent the standard errors of the means.

3.1.5.2. Tempo. While there was no significant difference in pleasantness ratings of music stimuli that varied in tempo ($Z = -0.12$, $P = 0.91$), the faster tempo music was significantly more stimulating than the slower tempo music ($Z = -2.10$, $P = 0.04$) (Table 2).

3.1.5.3. Volume. As shown in Table 2, the louder music stimulus was rated significantly more pleasant than the quieter music ($Z = -7.49$, $P < 0.001$). The louder music was also rated significantly more stimulating than the quieter music ($Z = -7.00$, $P < 0.001$).

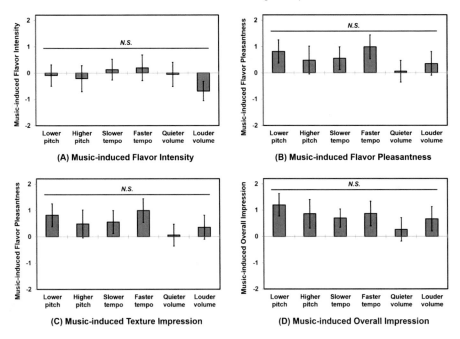

Figure 3. Mean ratings of music-induced flavor intensity (A), music-induced flavor pleasantness (B), music-induced texture impression (C), and music-induced overall impression (D) of bell peppers with respect to element levels of music (Study 1). For music-induced sensory perception or impression, the intensity or impression ratings of bell peppers under the silent condition were subtracted from those ratings in the presence of music stimulus (e.g., flavor intensity rating of bell peppers in the presence of higher-pitched music minus those of bell peppers under the silent condition). It should be noted that during each test session, participants evaluated bell peppers in a silent condition. Those subtracted values were then compared using the Friedman test. *N.S.* indicates no significance at $P < 0.05$. Error bars represent the standard errors of the means.

3.1.6. Relationships Between Individual Ratings of Music Pleasantness or Stimulation With Sensory Attribute and Impressions of Food Samples

3.1.6.1. Pitch. Spearman partial correlation analyses found that ratings of music pleasantness were significantly correlated with ratings of flavor pleasantness ($\rho = 0.13$, $P = 0.049$) and overall impression ($\rho = 0.13$, $P = 0.04$) (Table 3). In addition, ratings of music stimulation were significantly correlated with ratings of flavor intensity ($\rho = 0.13$, $P = 0.04$) and texture impression ($\rho = 0.14$, $P = 0.03$).

3.1.6.2. Tempo. Table 3 also shows the significant correlations of music pleasantness with flavor pleasantness ($\rho = 0.14$, $P = 0.03$), texture impression ($\rho = 0.17$, $P = 0.008$), and overall impression ($\rho = 0.15$, $P = 0.02$) of food samples. No significant relationship between ratings of music stimulation and flavor intensity was observed. In addition, individual ratings of music stimulation were significantly correlated with ratings of flavor intensity

Table 2.
Mean ratings ($\pm$ standard deviation) of pleasantness and stimulation for music stimuli arranged into two levels as a function of music elements: pitch, tempo, and volume[1]

	Levels of music element		
Pitch	Lower-pitched	Higher-pitched	P-value[2]
Music pleasantness	7.07 ($\pm$ 1.84)	5.66 ($\pm$ 2.34)	<0.001
Music stimulation	5.93 ($\pm$ 2.21)	5.24 ($\pm$ 2.33)	0.01
Tempo	Slower	Faster	P-value
Music pleasantness	7.33 ($\pm$ 1.57)	7.37 ($\pm$ 1.42)	0.91
Music stimulation	6.27 ($\pm$ 2.02)	6.56 ($\pm$ 1.99)	0.04
Volume	Quieter	Louder	P-value
Music pleasantness	5.71 ($\pm$ 1.59)	7.44 ($\pm$ 1.39)	<0.001
Music stimulation	5.02 ($\pm$ 1.66)	6.56 ($\pm$ 1.79)	<0.001

[1]Each participant rated pleasantness and stimulation for each music stimulus twice (i.e., during both sessions of chocolate and bell peppers).
[2]A significant difference was determined by the Wilcoxon signed-rank test at $P < 0.05$.

Table 3.
Spearman partial correlation coefficients (*P*-value) of music pleasantness and stimulation with flavor intensity, flavor pleasantness, texture impression, and overall impression of food samples

	Flavor intensity	Flavor pleasantness	Texture impression	Overall impression
Pitch				
Music pleasantness[1]	0.02 (0.78)	0.13 (0.049)	0.10 (0.11)	0.13 (0.04)
Music stimulation[2]	0.13 (0.04)	0.12 (0.06)	0.14 (0.03)	0.12 (0.06)
Tempo				
Music pleasantness	−0.02 (0.79)	0.14 (0.03)	0.17 (0.008)	0.15 (0.02)
Music stimulation	0.24 (<0.001)	0.16 (0.02)	0.20 (0.001)	0.16 (0.01)
Volume				
Music pleasantness	−0.02 (0.73)	0.08 (0.24)	0.05 (0.45)	0.13 (0.05)
Music stimulation	0.17 (<0.01)	0.08 (0.24)	0.10 (0.11)	0.06 (0.39)

[1]Individual ratings of music stimulation were used as covariance.
[2]Individual ratings of music pleasantness were used as covariance.

($\rho = 0.24$, $P < 0.001$), flavor pleasantness ($\rho = 0.16$, $P = 0.02$), texture impression ($\rho = 0.20$, $P = 0.001$), and overall impression ($\rho = 0.16$, $P = 0.01$).

3.1.6.3. Volume. When looking at correlations between individual ratings of music pleasantness or stimulation with flavor intensity, flavor pleasantness, texture impression, and overall impression of food samples, the only statistically significant correlation was between music stimulation and flavor intensity ($\rho = 0.17$, $P < 0.001$) (Table 3).

3.2. Study 2

3.2.1. Variations in Perceptions of Music Stimuli Arranged into Two Levels as a Function of Music Elements: Pitch, Tempo, and Volume

As shown in Fig. 4A, the lower-pitched and higher-pitched music stimuli differed significantly not only in the subjective perception of pitch level ($Z = -9.02$, $P < 0.001$), but also in the perceptions of tempo level ($Z = -4.54$, $P < 0.001$) and volume level ($Z = -7.66$, $P < 0.001$). The slower- and faster tempo music stimuli were found to differ significantly in the perception of tempo level ($Z = -8.60$, $P < 0.001$), as well as in the perception of volume level ($Z = -2.53$, $P = 0.01$); there was no significant difference between the two music stimuli with respect to the perception of pitch level ($Z = -1.37$, $P = 0.17$) (Fig. 4B). Additionally, as shown in Fig. 4C, the quieter and louder music stimuli differed significantly not only in the perception of volume level ($Z = -9.70$, $P < 0.001$), but also in the perceptions of pitch level ($Z = -4.28$, $P < 0.001$) and tempo level ($Z = -6.27$, $P < 0.001$).

3.2.2. Variations in Ratings of Pleasantness and Stimulation for Music Stimuli Arranged into Two Levels as a Function of Music Elements: Pitch, Tempo, and Volume

There were wide ranges in the individual ratings of pleasantness or stimulation for music stimuli arranged into the two levels (i.e., lower and upper) with respect to pitch, tempo, or volume (Fig. 4). Figure 4A shows that, while there was no significant difference between the lower-pitched and higher-pitched music stimuli in terms of stimulation ratings of music stimuli ($Z = -0.12$, $P = 0.91$), the lower-pitched music stimulus was rated more pleasant than the higher-pitched music stimulus ($Z = -7.42$, $P < 0.001$). Figure 4B demonstrates that, even though no significant difference between the slower tempo and faster tempo music stimuli was observed with respect to stimulation ratings of music stimuli ($Z = -0.18$, $P = 0.86$), the slower tempo music stimulus was rated more pleasant than the faster tempo music stimulus ($Z = -5.83$, $P < 0.001$). Finally, there were significant differences between the quieter and louder music stimuli with respect to ratings of music pleasantness ($Z = -2.75$, $P = 0.006$) and stimulation ($Z = -6.92$, $P < 0.001$). More specifically, the louder music stimulus was rated more pleasant and stimulating than the quieter music stimulus (Fig. 4C).

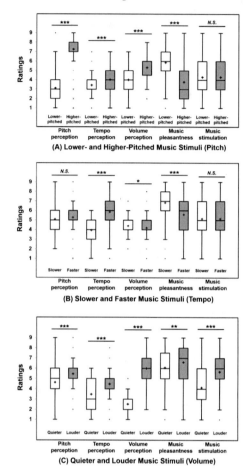

Figure 4. Comparisons between the two music stimuli arranged into two levels of pitch (A), tempo (B), or volume (C) with respect to subjective ratings of pitch perception, tempo perception, volume perception, music pleasantness, and music stimulation (Study 2). In each box plot, the cross and the central horizontal bar represent the mean and the median, respectively. The lower and upper limits of the box indicate the first and third quartiles, respectively. Points below or above the whisker's lower and upper bounds represent outliers. *N.S.* indicates no significance at $P < 0.05$. *, **, and *** indicate significance at $P < 0.05$, $P < 0.01$, and $P < 0.001$, respectively.

4. Discussion

4.1. Effects of Musical Pitch Levels on Food Perception and Acceptance

This study showed small but significant effects of musical pitch levels on texture impression of chocolate (Fig. 1A) and overall acceptance of bell peppers (Fig. 1B). Texture of chocolate was rated as more pleasant in the presence of lower-pitched music than in the presence of higher-pitched music. Overall impression of bell peppers was also rated significantly higher in the presence of lower-pitched music than under a silent condition. These results are opposite of what we expected. It has been shown that high-pitched stimuli are considered to increase positive and active emotions (Balkwill and Thompson, 1999; Collier and Hubbard, 1998; Hevner, 1937), thereby increasing hedonic ratings of food samples (Kantono *et al.*, 2016b, c). However, it is worth noting that the low-pitched stimuli have been also found to show positive associations with pleasantness (Ilie and Thompson, 2006; Scherer and Oshinsky, 1977) or arousal (Jaquet *et al.*, 2014). In other words, pleasantness and arousal evoked by pitch-variable stimuli may vary depending on the sound contexts, i.e., the combination and interaction with other structural elements (Gabrielsson and Lindström, 2001) and personal factors, such as gender (Jaquet *et al.*, 2014). Moreover, it should be noted that the pleasantness and stimulation ratings of the lower-pitched music alone were both higher than those of the higher-pitched music in this study (Table 2). In this sense, pleasantness and arousal evoked by a lower-pitched music stimulus might increase texture impression of chocolate and overall impression of bell peppers. This assumption was supported by the Spearman partial correlation analyses summarized in Table 3. Specifically, both music pleasantness and stimulation are positively correlated with hedonic ratings of food samples, suggesting that pleasant and stimulating emotions evoked by music stimuli varying in pitch levels may be associated with elevated likings of food samples. Hedonic halo effects of music stimuli on subsequently or simultaneously administered stimuli have been also observed in other studies (Fiegel *et al.*, 2014; Kantono *et al.*, 2016a, b, c, 2018; Reinoso Carvalho *et al.*, 2016; Seo and Hummel, 2011; Seo *et al.*, 2014). For example, Kantono *et al.* (2016b) found that liked music stimuli, in comparison to disliked music stimuli, increased hedonic impression of chocolate gelati. In addition, Reinoso Carvalho *et al.* (2016) found positive correlations between soundtrack liking and hedonic impression of beer. Similarly, Fiegel *et al.* (2014) also showed positive correlations of hedonic impression of food samples with pleasantness and stimulation ratings of music stimuli. However, a further study is needed to fully determine whether hedonic halo effects of musical stimuli can vary as a function of taste quality or food item.

Previous research has found that sweetness and sourness correspond with higher-pitched sounds, while bitterness corresponds with lower-pitched sounds (Crisinel and Spence, 2009, 2010; Knoeferle *et al.*, 2015; Reinoso Carvalho *et al.*, 2016; Wang *et al.*, 2015; for a review, Knöferle and Spence, 2012; also see Höchenberger and Ohla, 2019). It has also been shown that perceived sweetness intensity increases with a higher-pitched stimulus (Crisinel and Spence, 2010; Mesz *et al.*, 2011). Considering these results, we initially expected that flavor intensity ratings of a sweet-tasting substance (milk chocolate) would be higher under the higher-pitched music, but there was no significant result. A plausible explanation for the lack of significance is that the higher-pitched music stimulus used in this study consists of a wide frequency range (880.0 Hz to 2093.0 Hz), which includes some frequencies that individuals might consider as relatively low-pitched (also see Drayna, 2007; Oxenham *et al.*, 2011). In other words, there was not a distinct classification between the low and high pitches of the music stimuli used. However, it should be noted that the auditory pitch-based cross-modal correspondences are more likely to be associated with relative nature of pitch (i.e., relatively lower or higher pitches) than absolute nature of pitch (i.e., specific pitch levels) (see Spence, 2019, for a review). It is also worth noting that pitch perception is affected by other music elements (e.g., timbre and loudness; Melara and Marks, 1990) as well as individual factors (Pitt, 1994). As displayed in Fig. 4A, while most participants in Study 2 rated the higher-pitched music stimulus as high-pitched (more than the sixth point on a nine-point scale), some participants perceived it as low-pitched (below the fifth point). Notably, subjective perceptions of tempo and volume were also affected by music stimuli arranged into two levels with respect to musical pitch. Furthermore, it is interesting to note that there was a significant correlation between individual ratings of music stimulation and flavor intensity (Table 3), suggesting that stimulating emotion evoked by music stimuli varying in pitch level may be associated with flavor intensity.

4.2. *Effects of Musical-Tempo Levels on Food Perception and Acceptance*

This study showed no significant effects of musical-tempo levels on flavor intensity and other hedonic ratings for the chocolate sample (Fig. 1C). Since the two stimuli were not significantly different in terms of music pleasantness (Table 2), music-evoked emotions might differ minimally, resulting in no significant difference among the musical-tempo conditions with respect to hedonic ratings of the chocolate sample. An acceptable tempo range for individuals is not distinguished because it depends on their age, culture, and other demographic characteristics (Milliman, 1986). Such individual variations with respect to the acceptable tempo range may lessen the effects of the tempo stimuli on evoked emotions, thus offering an explanation for no significant effect of musical-tempo levels on consumer behaviors (Caldwell and Hibbert, 2002).

The faster tempo music stimulus was found to increase flavor pleasantness of bell peppers when compared to the silent condition. Furthermore, texture of bell peppers was rated as more pleasant in both the faster and slower tempo music conditions compared to the silent condition (Fig. 4B). These results suggest that regardless of tempo, a music stimulus can affect hedonic impression of bell peppers, most notably the texture impression (also see Sullivan, 2002).

It has been shown that passages presented in a fast tempo tend to elicit feelings of happiness and pleasantness, while slow tempo stimuli produce more tranquil, sentimental, and solemn feelings (Collier and Hubbard, 1998; Hevner, 1937). While our findings are consistent with these assumptions, the positive ratings associated with fast tempo music may also be explained by the act of transferring positive feelings from one stimulus to another subsequent or simultaneous stimulus (Logeswaran and Bhattacharya, 2009; Seo and Hummel, 2011; Woods *et al.*, 2011). This effect is visible in correlations of music pleasantness with flavor pleasantness, texture impression, and overall impression of food samples (Table 3). In addition, music stimulation was found to correlate with flavor pleasantness, texture impression, and overall impression of food samples (Table 3). Similar trends were also observed in other studies (Fiegel *et al.*, 2014; Kantono *et al.*, 2016a, b, c; Wang *et al.*, 2016). Although flavor intensity of food samples did not differ between the two musical-tempo conditions, it might be related to stimulating emotion elicited by music stimuli varying in tempo. More specifically, music stimulation showed a significantly positive correlation with flavor intensity of food samples (Table 3).

4.3. Effects of Musical Volume Levels on Food Perception and Acceptance

Loud stimuli are associated with liveliness, energy, and agitation, while quiet stimuli are usually peaceful, gentle, and dreamy (Berger and Schneck, 2003; Cooke, 1962). When a volume range of a sound stimulus is rated as acceptable by a listener, the sound stimulus causes the listener to evaluate a test sample as more positive, thus increasing his/her overall satisfaction of the sample (Novak *et al.*, 2010). However, a stimulus exceeding the volume zone of tolerance creates dissatisfaction and mental stress (Beverland *et al.*, 2006; Ferber and Cabanac, 1987; Novak *et al.*, 2010). In our study, the louder music stimulus was evaluated as more pleasant than the quieter stimulus (Table 2), possibly attributable to the significantly higher ratings of flavor pleasantness, texture impression, and overall impression of chocolate sample in the presence of louder volume music (Fig. 1E). In previous research, sound amplitude was directly related to food stimulus quality (Woods *et al.*, 2011), pleasantness (Zampini and Spence, 2004), and increased liking (Spence and Shankar, 2010). Considering these collective effects, there is sufficient evidence to support our findings of chocolate flavor pleasantness, texture impression, and overall impression being rated highest under a loud volume condition.

Although sweet items have exhibited higher (Ferber and Cabanac, 1987; Stafford *et al.*, 2012) or lower (Woods *et al.*, 2011; Yan and Dando, 2015) intensities of sweetness when consumed in the presence of loud background noise or music stimuli, flavor intensity was the only attribute for chocolate that was not found to be significantly different between the loud and quiet music stimuli (Fig. 1E). Since sweet taste is only a part of 'flavor' constituents, further study is needed to determine whether the louder music stimulus can solely affect sweetness intensity of chocolate. In addition, since there was a large individual variation in the subjective perception of louder volume music (Fig. 4C), it would be interesting to examine whether further louder music stimuli can alter either sweetness or flavor intensity of tasting substances.

People tend to consume sweet and fatty foods to alleviate stress (Gibson, 2006) exhibiting a strong correlation of emotional eating with the consumption of energy dense and sweet foods, in contrast to the consumption of vegetables and fruits (Konttinen *et al.*, 2010). This trend suggests that sweet foods (e.g., 'emotional foods') are more likely to alter and mediate emotions than non-emotional foods, such as bell peppers. In our study, for bell peppers, flavor intensity was the only attribute showing a significant difference between the quieter and louder music stimuli, while a post-hoc test revealed no significant effect of musical volume levels on flavor intensity (Fig. 1F). In addition, because bell peppers produce a greater level of noise stemming from the mastication process than chocolate, the mastication process-induced noise might lessen the loud music-evoked emotions, resulting in no significant impact of musical volume levels on hedonic ratings of bell peppers. Similar results involving mastication noise have been also observed in previous studies (Fiegel *et al.*, 2014; Herrington, 1996; for a review, Spence, 2015).

4.4. Variations in Music-Induced Sensory Perception and Impression of Food Samples as a Function of Music Element Levels

This study determined whether and how music-induced flavor intensity, flavor pleasantness, texture impression, and overall impression of food samples could differ among all six music stimuli varying in pitch, tempo, or volume levels. As shown in Fig. 2, the louder music stimulus increased hedonic ratings of chocolate flavor significantly more than the higher-pitched or slower tempo music stimulus. In addition, the louder music stimulus increased acceptance of chocolate texture significantly more than the higher-pitched, faster tempo, or quieter music stimulus. These results suggest that individuals are likely to enjoy the flavor and texture of chocolate more when they turn up the volume of the music stimulus. However, no significant differences among the six music stimulus conditions were observed with respect to the music-induced flavor intensity and overall impression of chocolate.

In contrast to chocolate, bell peppers did not show significant differences among the six music stimulus conditions with respect to music-induced flavor intensity, flavor pleasantness, texture impression, and overall impression. As mentioned above, no significant effect of music element levels on sensory perception and impression of bell peppers might be related to 1) their non-emotional food characteristics and 2) loud noises elicited by mastication process during consumption.

4.5. *Limitations and Future Studies*

Although our findings present some empirical evidence that element levels of music can affect hedonic ratings of food samples, it is important to consider all the stimuli and environmental characteristics within which the study was performed. Firstly, this study was carried out in sensory booths using pieces of chocolate and bell peppers. The emotional influence or perceptions formed might differ if the same study was performed in realistic contexts such as at a restaurant or home, because different atmospherics may lead to different mood states resulting in different behavioral responses (Herrington, 1996; Kantono *et al.*, 2018). Furthermore, associations between music and food samples might differ if the same study was conducted using different types of food samples (e.g., pronounced in sour, salty, or bitter taste). Secondly, a further study with a larger spectrum of participants' demographics may be needed to generalize the present findings because only young adults (95% Caucasians; 75% females) who live in Northwest Arkansas participated in this study (Henrich *et al.*, 2010). For example, since extroverts need higher levels of stimulation in comparison to introverts (Eysenck, 1967), if a population contains more extroverts, the perception and emotion may be heavily weighted to one side due to participants' personalities. Finally, it would be interesting to explore whether the effects of music element levels on sensory perception and hedonic impression of foods consumed can be observed in music genres other than classical, as Eerola (2011) showed that emotions can function differently across genres.

5. Conclusion

Our findings suggest that the effect of background music on sensory perception and hedonic impression of food samples can differ with respect to the upper/lower level of music elements such as pitch, tempo, or volume. In addition, pleasantness (valence) and stimulation (arousal) emotions evoked by music stimuli varying in structural element levels appear to play an important role in the effect of background music on sensory perception and acceptance of foods. However, it should also be noted that the influences of music element levels were food-dependent, and there were large individual variations in the music-evoked pleasantness and stimulation, thereby lessening the impact of element levels on sensory perception and acceptance of foods.

References

Addinsoft (2019). XLSTAT statistical and data analysis solution. https://www.xlstat.com. Retrieved 14 January 2019.

Balkwill, L. and Thompson, W. (1999). A cross-cultural investigation of the perception of emotion in music: psychophysical and cultural cues, *Music Percept.* **17**, 43–64.

Berger, D. S. and Schneck, D. J. (2003). The use of music therapy as a clinical intervention for physiologic functional adaptation, *J. Sci. Explor.* **17**, 687–703.

Beverland, M., Lim, E. A. C., Morrison, M. and Terziovski, M. (2006). In store music and consumer–brand relationships: relational transformation following experiences of (mis)fit, *J. Bus. Res.* **59**, 982–989.

Biswas, D., Lund, K. and Szocs, C. (2019). Sounds like a healthy retail atmospheric strategy: effects of ambient music and background noise on food sales, *J. Acad. Mark. Sci.* **47**, 37–55.

Bradley, M. M. and Lang, P. J. (1994). Measuring emotions: the self-assessment manikin and the semantic differential, *J. Behav. Ther. Exp. Psychiatry* **25**, 49–59.

Bruner, C. G. (1990). Music, mood, and marketing, *J. Mark.* **54**, 94–104.

Burton, R. (2015). The elements of music: what are they, and who cares?, in: *Music: Educating for Life. ASME XXth National Conference Proceedings*, pp. 22–28. Australian Society for Music Education, Parkvill, VIC, Australia.

Caldwell, C. and Hibbert, S. A. (2002). The influence of music tempo and musical preference on restaurant patrons' behavior, *Psychol. Mark.* **19**, 895–917.

Collier, W. G. and Hubbard, T. L. (1998). Judgements of happiness, brightness, speed and tempo change of auditory stimuli varying in pitch and tempo, *Psychomusicology* **17**, 36–55.

Cooke, D. (1962). *The Language of Music*. Oxford University Press, London, UK.

Crisinel, A.-S. and Spence, C. (2009). Implicit association between basic tastes and pitch, *Neurosci. Lett.* **464**, 39–42.

Crisinel, A.-S. and Spence, C. (2010). As bitter as a trombone: synesthetic correspondences in nonsynestheses between tastes/flavors and musical notes, *Atten. Percept. Psychophys.* **72**, 1994–2002.

Donovan, R. J. and Rossiter, J. R. (1982). Store atmosphere: an environmental psychology approach, *J. Retail.* **58**, 34–57.

Drayna, D. T. (2007). Absolute pitch: a special group of ears, *Proc. Natl Acad. Sci. USA* **104**, 14549–14550.

Eerola, T. (2011). Are the emotions expressed in music genre-specific? An audio-based evaluation of datasets spanning classical, film, pop and mixed genres, *J. New Music Res.* **40**, 349–366.

Engels, R. C. M. E., Poelen, E. A. P., Spijkerman, R. and Bogt, T. T. (2012). The effects of music genre on young people's alcohol consumption: an experimental observational study, *Subst. Use Misuse* **47**, 180–188.

Eysenck, H. J. (1967). *The Biological Basis of Personality*. Charles Thomas, Springfield, IL, USA.

Ferber, C. and Cabanac, M. (1987). Influence of noise on gustatory affective ratings and preference for sweet or salt, *Appetite* **8**, 229–235.

Fiegel, A., Meullenet, J.-F., Harrington, R. J., Humble, R. and Seo, H.-S. (2014). Background music genre can modulate flavor pleasantness and overall impression of food stimuli, *Appetite* **76**, 144–152.

Friedman, M. (1937). The use of ranks to avoid the assumption of normality implicit in the analysis of variance, *J. Am. Stat. Assoc.* **32**, 675–701.

Gabrielsson, A. and Lindström, E. (2001). The influence of musical structure on emotional expression, in: *Music and Emotion: Theory and Research*, P. N. Juslin and J. A. Sloboda (Eds), pp. 223–248. Oxford University Press, Oxford, UK.

Gacula Jr., M. and Rutenbeck, S. (2006). Sample size in consumer test and descriptive analysis, *J. Sens. Stud.* **21**, 129–145.

Garlin, F. V. and Owen, K. (2006). Setting the tone with the tune: a meta-analytic review of the effects of background music in retail settings, *J. Bus. Res.* **59**, 755–764.

Gibson, E. L. (2006). Emotional influences on food choice: sensory, physiological and psychological pathways, *Physiol. Behav.* **89**, 53–61.

Gomez, P. and Danuser, B. (2007). Relationships between musical structure and psychophysiological measures of emotion, *Emotion* **7**, 377–387.

Guéguen, N., Le Guellec, H. and Jacob, C. (2004). Sound level of background music and alcohol consumption. an empirical evaluation, *Percept. Mot. Skills* **99**, 34–38.

Guéguen, N., Jacob, C., Le Guellec, H., Morineau, T. and Lourel, M. (2008). Sound level of environmental music and drinking behavior: a field experiment with beer drinkers, *Alcohol. Clin. Exp. Res.* **32**, 1795–1798.

Henrich, J., Heine, S. J. and Norenzayan, A. (2010). The weirdest people in the world?, *Behav. Brain Sci.* **33**, 61–83.

Herrington, J. D. (1996). Effects of music in service environments: a field study, *J. Serv. Mark.* **10**, 26–41.

Hevner, K. (1937). The affective value of pitch and tempo in music, *Am. J. Psychol.* **49**, 621–630.

Höchenberger, R. and Ohla, K. (2019). A bittersweet symphony: evidence for taste-sound correspondences without effects on taste quality-specific perception, *J. Neurosci. Res.* **97**, 267–275.

Ilie, G. and Thompson, W. F. (2006). A comparison of acoustic cues in music and speech for three dimensions of affect, *Music Percept.* **23**, 319–330.

Jacob, C. (2006). Styles of background music and consumption in a bar. An empirical evaluation, *Int. J. Hosp. Manag.* **25**, 716–720.

Jaquet, L., Danuser, B. and Gomez, P. (2014). Music and felt emotions: how systemic pitch level variations affect the experience of pleasantness and arousal, *Psychol. Music* **42**, 51–70.

Kantono, K., Hamid, N., Shepherd, D., Lin, Y. H. T., Yakuncheva, S., Yoo, M. J. Y., Grazioli, G. and Carr, B. T. (2016a). The influence of auditory and visual stimuli on the pleasantness of chocolate gelati, *Food Qual. Pref.* **53**, 9–18.

Kantono, K., Hamid, N., Shepherd, D., Yoo, M. J. Y., Carr, B. T. and Grazioli, G. (2016b). The effect of background music on food pleasantness ratings, *Psychol. Music* **44**, 1111–1125.

Kantono, K., Hamid, N., Shepherd, D., Yoo, M. J. Y., Grazioli, G. and Carr, B. T. (2016c). Listening to music can influence hedonic and sensory perceptions of gelati, *Appetite* **100**, 244–255.

Kantono, K., Hamid, N., Shepherd, D., Lin, Y. H. T., Brad, C., Grazioli, G. and Carr, B. T. (2018). The effect of music on gelato perception in different eating contexts, *Food Res. Int.* **113**, 43–56.

Kendall, M. G. and Smith, B. B. (1939). The problem of *m* rankings, *Ann. Math. Stat.* **10**, 275–287.

Khalfa, S., Roy, M., Rainville, P., Dalla Bella, S. and Peretz, I. (2008). Role of tempo entrainment in psychophysiological differentiation of happy and sad music?, *Int. J. Psychophysiol.*
68, 17–26.

Knöferle, K. and Spence, C. (2012). Crossmodal correspondences between sounds and tastes,
Psychon. Bull. Rev. **19**, 992–1006.

Knoeferle, K. M., Woods, A., Käppler, F. and Spence, C. (2015). That sounds sweet: using
cross-modal correspondences to communicate gustatory attributes, *Psychol. Mark.* **32**, 107–
120.

Konttinen, H., Männistö, S., Sarlio-Lähteenkorva, S., Silventoinen, K. and Haukkala, A. (2010).
Emotional eating, depressive symptoms and self-reported food consumption. A population-
based study, *Appetite* **54**, 473–479.

Logeswaran, N. and Bhattacharya, J. (2009). Crossmodal transfer of emotion by music, *Neurosci. Lett.* **455**, 129–133.

Mammasse, N. and Schlich, P. (2014). Adequate number of consumers in a liking test. Insights
from resampling in seven studies, *Food Qual. Pref.* **31**, 124–128.

McElrea, H. and Standing, L. (1992). Fast music causes fast drinking, *Percept. Mot. Skills* **75**,
362.

Mehrabian, A. and Russell, J. A. (1974). *An Approach to Environmental Psychology.* MIT Press,
Cambridge, MA, USA.

Melara, R. D. and Marks, L. E. (1990). Interaction among auditory dimensions: timbre, pitch,
and loudness, *Percept. Psychophys.* **48**, 169–178.

Mesz, B., Trevisan, M. A. and Sigman, M. (2011). The taste of music, *Perception* **40**, 209–219.

Milliman, R. E. (1982). Using background music to affect the behavior of supermarket shoppers,
J. Mark. **46**, 86–91.

Milliman, R. E. (1986). The influence of background music on the behavior of restaurant patrons, *J. Consum. Res.* **13**, 286–289.

Moskowitz, H. R. (1997). Base size in product testing: a psychophysical viewpoint and analysis,
Food Qual. Pref. **8**, 247–255.

Muniz, R., Harrington, R. J., Ogbeidea, G.-C. and Seo, H.-S. (2017). The role of sound congruency on ethnic menu item selection and price expectations, *Int. J. Hosp. Tour. Admin.* **18**,
245–271.

Noel, C. and Dando, R. (2015). The effect of emotional state on taste perception, *Appetite* **95**,
89–95.

Novak, C. C., La Lopa, J. and Novak, R. E. (2010). Effects of sound pressure levels and sensitivity to noise on mood and behavioral intent in a controlled fine dining restaurant environment,
J. Culin. Sci. Technol. **8**, 191–218.

Oxenham, A. J., Micheyl, C., Keebler, M. V., Loper, A. and Santurette, S. (2011). Pitch perception beyond the traditional existence region of pitch, *Proc. Natl Acad. Sci. USA* **108**,
7629–7634.

Pellegrino, R., Luckett, C. R., Shinn, S. E., Mayfield, S., Gude, K., Rhea, A. and Seo, H.-S.
(2015). Effects of background sound on consumers' sensory discriminatory ability among
foods, *Food Qual. Pref.* **43**, 71–78.

Pitt, M. A. (1994). Perception of pitch and timbre by musically trained and untrained listeners,
J. Exp. Psychol. Hum. Percept. Perform. **20**, 976–986.

Pollatos, O., Kopietz, R., Linn, J., Albrecht, J., Sakar, V., Anzinger, A., Schandry, R. and Wiesmann, M. (2007). Emotional stimulation alters olfactory sensitivity and odor judgment, *Chem. Sens.* **32**, 583–589.

Reinoso Carvalho, F., Wang, Q. (J.), Van Ee, R. and Spence, C. (2016). The influence of soundscapes on the perception and evaluation of beers, *Food Qual. Pref.* **52**, 32–41.

Reinoso Carvalho, F., Wang, Q. (J.), Van Ee, R., Persoone, D. and Spence, C. (2017). "Smooth operator": music modulates the perceived creaminess, sweetness, and bitterness of chocolate, *Appetite* **108**, 383–390.

Roballey, T. C., McGreevy, C., Rongo, R. R., Schwantes, M. L., Steger, P. J., Wininger, M. A. and Gardner, E. B. (1985). The effect of music on eating behavior, *Bull. Psychon. Soc.* **23**, 221–222.

Scherer, K. R. and Oshinsky, J. S. (1977). Cue utilization in emotion attribution from auditory stimuli, *Motiv. Emot.* **1**, 331–346.

Seo, H.-S. and Hummel, T. (2011). Auditory–olfactory integration: congruent or pleasant sounds amplify odor pleasantness, *Chem. Sens.* **36**, 301–309.

Seo, H.-S., Lohse, F., Luckett, C. R. and Hummel, T. (2014). Congruent sound can modulate odor pleasantness, *Chem. Sens.* **39**, 215–228.

Septianto, F. (2016). "Chopin" effect? An exploratory study on how musical tempo influence consumer choice of drink with different temperatures, *Asia Pac. J. Mark. Logist.* **28**, 765–779.

Shapiro, S. S. and Wilk, M. B. (1965). An analysis of variance test for normality (complete samples), *Biometrika* **52**, 591–611.

Spence, C. (2012). Auditory contributions to flavour perception and feeding behaviour, *Physiol. Behav.* **107**, 505–515.

Spence, C. (2014). Noise and its impact on the perception of food and drink, *Flavour* **3**, 9. DOI:10.1186/2044-7248-3-9.

Spence, C. (2015). Eating with our ears: assessing the importance of the sounds of consumption on our perception and enjoyment of multisensory flavour experiences, *Flavour* **4**, 3. DOI:10.1186/2044-7248-4-3.

Spence, C. (2019). On the relative nature of (pitch-based) crossmodal correspondences, *Multisens. Res.* **32**, 235–265. DOI:10.1163/22134808-20191407.

Spence, C. and Shankar, M. U. (2010). The influence of auditory cues on the perception of, and responses to, food and drink, *J. Sens. Stud.* **25**, 406–430.

Spence, C., Puccinelli, N. M., Grewal, D. and Roggeveen, A. L. (2014). Store atmospherics: a multisensory perspective, *Psychol. Mark.* **31**, 472–488.

Stafford, L. D., Fernandes, M. and Agobiani, E. (2012). Effects of noise and distraction on alcohol perception, *Food Qual. Pref.* **24**, 218–224.

Stroebele, N. and de Castro, J. M. (2006). Listening to music while eating is related to increases in people's food intake and meal duration, *Appetite* **47**, 285–289.

Sullivan, M. (2002). The impact of pitch, volume and tempo on the atmospheric effects of music, *Int. J. Retail Distrib. Manag.* **30**, 323–330.

Tomczak, M. and Tomczak, E. (2014). The need to report effect size estimates revisited. An overview of some recommended measures of effect size, *Trends Sport Sci.* **1**, 19–25.

Vickers, Z. M. (1991). Sound perception and food quality, *J. Food Qual.* **14**, 87–96.

Wang, Q. (J.), Woods, A. and Spence, C. (2015). "What's your taste in music?" A comparison of the effectiveness of various soundscapes in evoking specific tastes, *i-Perception* **6**, 2041669515622001. DOI:10.1177/2041669515622001.

Wang, Q. J., Wang, S. and Spence, C. (2016). "Turn up the taste": assessing the role of taste intensity and emotion in mediating crossmodal correspondences between basic tastes and pitch, *Chem. Sens.* **41**, 345–356.

Webster, G. and Weir, C. (2005). Emotional responses to music: interactive effects of mode, texture, and tempo, *Motiv. Emot.* **29**, 19–39.

Williams, E. J. (1949). Experimental designs balanced for the estimation of residual effects of treatments, *Aust. J. Sci. Res.* **A2**, 149–168.

Wilson, S. (2003). The effect of music on perceived atmosphere and purchase intentions in a restaurant, *Psychol. Music* **31**, 91–112.

Woods, A. T., Poliakoff, E., Lloyd, D. M., Kuenzel, J., Hodson, R., Gonda, H., Batchelor, J., Dijksterhuis, G. B. and Thomas, A. (2011). Effect of background noise on food perception, *Food Qual. Pref.* **22**, 42–47.

Yalch, R. and Spangenberg, E. (1990). Effects of store music on shopping behavior, *J. Consum. Mark.* **7**, 55–63.

Yan, K. S. and Dando, R. (2015). A crossmodal role for audition in taste perception, *J. Exp. Psychol. Hum. Percept. Perform.* **41**, 590–596.

Yang, M. and Kang, J. (2016). Pitch features of environmental sounds, *J. Sound Vib.* **374**, 312–328.

Zampini, M. and Spence, C. (2004). The role of auditory cues in modulating the perceived crispness and staleness of potato chips, *J. Sens. Stud.* **19**, 347–363.

Zampini, M. and Spence, C. (2005). Modifying the multisensory perception of a carbonated beverage using auditory cues, *Food Qual. Pref.* **16**, 632–641.

Zellner, D., Geller, T., Lyons, S., Pyper, A. and Riaz, K. (2017). Ethnic congruence of music and food affects food selection but not liking, *Food Qual. Pref.* **56**, Part A, 126–129.

High-Tempo and Stinky: High Arousal Sound–Odor Congruence Affects Product Memory

Marijn Peters Rit, Ilja Croijmans [1,*] **and Laura J. Speed** [2,*,**]

[1] Faculty of Social and Behavioural Sciences, Utrecht University, The Netherlands
[2] Department of Psychology, University of York, UK

Abstract

The tendency to match different sensory modalities together can be beneficial for marketing. Here we assessed the effect of sound–odor congruence on people's attitude and memory for products of a familiar and unfamiliar brand. Participants smelled high- and low-arousal odors and then saw an advertisement for a product of a familiar or unfamiliar brand, paired with a high- or low-arousal jingle. Participants' attitude towards the advertisement, the advertised product, and the product's brand was measured, as well as memory for the product. In general, no sound–odor congruence effect was found on attitude, irrespective of brand familiarity. However, congruence was found to affect recognition: when a high-arousal odor and a high-arousal sound were combined, participants recognized products faster than in the other conditions. In addition, familiar brands were recognized faster than unfamiliar brands, but only when sound or odor arousal was high. This study provides insight into the possible applications of sound–odor congruence for marketing by demonstrating its potential to influence product memory.

Keywords

Crossmodal congruence, arousal, attitude, product memory, brand, sound, odor

1. Introduction

Whether it is a bottle of cola, a request for charity money, or a picture portraying the perfect car, we are exposed to various forms of advertising every day. Advertisements have become part of our daily lives (Dyer, 1982). Because of technological developments like the internet and social media, one

---

[*] Authors contributed equally to this work.
[**] To whom correspondence should be addressed. E-mail: laura.speed@york.ac.uk

© KONINKLIJKE BRILL NV, LEIDEN, 2019 | DOI:10.1163/9789004416307_006

can send and receive more messages than ever before, which can be more information than people can take in (Cheng *et al.*, 2010). Partly because of this information overload, we have developed methods, such as cognitive avoidance, to counter our exposure to advertisements (Fransen *et al.*, 2015). In turn, advertisement companies must develop adverts that stand out and will be remembered. Crossmodal correspondences pose new and exciting possibilities to achieve this (Velasco and Spence, 2019).

Our senses constantly deal with multiple signals. We need to determine which of the diverse set of signals belong to the same object, whilst separating the signals that do not (Treisman, 1996, 1998). Crossmodal correspondences can develop when some sensory combinations are more salient than others (Simner *et al.*, 2010), and they are most likely to manifest themselves when perceptual stimuli are presented in close temporal proximity (Spence, 2011). In other words, crossmodal correspondence is "a tendency for a sensory feature, or attribute, in one modality, either physically present or merely imagined, to be matched (or associated) with a sensory feature in another sensory modality" (Spence and Parise, 2012, p. 410).

Crossmodal correspondences can help us to make sense of our environment (Spence, 2011) and can influence people's perception and behavior (e.g., see Piqueras-Fiszman and Spence, 2015), making them important for marketing and consumer research. In particular, *congruence* in crossmodal correspondences may be an efficient marketing attribute. For example, congruence between odor and touch in terms of gender and temperature associations led to positive evaluations (Krishna et al., 2010), and in terms of marketing-relevant behavior, it has been shown that customers are more likely to buy products from a shelf that has a color congruent with the frequency of music being played (Hagtvedt and Brasel, 2016). Congruence can lead to positive responses because people like experiences that are expected or predictable (Mandler, 1982, cited in Meyers-Levy and Tybout, 1989). This paper investigates congruence between two modalities that are relevant for marketers but are often overlooked: audition and olfaction. Olfaction is a relatively novel domain, seldom utilized in marketing contexts. However, it is crucially involved in flavor perception (Auvray and Spence, 2008; Small and Prescott, 2005), and is thus important when considering multisensory food marketing. Despite the potential utility of olfaction in adverts, odors and sounds have not frequently been investigated together.

Radio and television adverts use sound to increase the appeal of a product or service, influencing behavior, mood and affect (Kellaris and Kent, 1993). According to Gabrielsson (2001), musical stimuli are one of the most powerful triggers of emotion (as cited in Rickard, 2004). When used in an advertisement, music can change how a consumer feels (Morris and Boone, 1998). For example, people's perceived arousal correlates with the intensity of a sound

(Mikutta *et al.*, 2013). Music also influences behavior: congruence between music in stores and the products on offer can influence purchase behavior (North *et al.*, 1999; Zellner *et al.*, 2017), and volume of music can affect sales of healthy versus unhealthy foods (Biswas *et al.*, 2019). Music can even influence people's taste perceptions (Fiegel *et al.*, 2014; Reinoso Carvalho *et al.*, 2015; Wang and Spence, 2016). Together, this illustrates music is a powerful tool to influence people's internal states and behavior. Stores widely apply this by creating a better shopping atmosphere with sounds (Petruzzellis *et al.*, 2014). In a similar vein, music-to-wine pairings are increasingly being used in multisensory tasting events, with music revolutionizing how wine is enjoyed by consumers (Spence, 2019). This underscores the role sound can play in crossmodal marketing for products where olfaction plays a paramount role, such as wine.

Like music, smell is also closely linked to emotion (Yeshurun and Sobel, 2010), with the olfactory bulb directly linked with the limbic system, thought to control basic emotions (Bradford and Desrochers, 2009). Thus, smell might also be used to influence behavior in a similar manner (e.g., de Lange *et al.*, 2012; Holland *et al.*, 2005; Liljenquist *et al.*, 2010). For example, the odor of a cleaning product can lead to more cleaning-related behavior (de Lange *et al.*, 2012; Holland *et al.*, 2005). Similarly, congruence between a product and its odor (e.g., chocolate and chocolate odor) can encourage people to spend more time choosing a product and thus process product information in depth, instead of applying simple heuristics (Mitchell *et al.*, 1995). Companies have begun using odor in these ways: take for example the smell of car interiors and freshly baked cookies in a bakery (Bradford and Desrochers, 2009; Moeran, 2007), although the effects found are not always in the direction predicted (De Wijk *et al.*, 2018), and too much odor exposure may even reduce purchase behavior (Biswas and Szocs, 2019).

How then, can sound and odor be combined to increase the effectiveness of advertisements? Previous research has suggested odor and sound are associated on certain dimensions. People have been shown to consistently match odors with specific pitches (Belkin *et al.*, 1997; Crisinel *et al.*, 2013; Deroy *et al.*, 2013), for example, caramel to low pitch (Belkin *et al.*, 1997). However, pitch can be difficult to control experimentally, since it is a subjective property varying in both loudness and tone (frequency). More importantly, the jingles or songs of advertisements are composed of various pitches, making it difficult to implement consistent, ecologically-valid odor–pitch associations within them (but see Crisinel *et al.*, 2012). Odors can also be congruent with sounds based on the sounds associated with the product (e.g., the smell of crisps and the sound of crisps being eaten; Seo and Hummel, 2011). However, such a manipulation may be obvious to participants, making the purpose of the experiment explicit. An alternative type of sound stimuli are musical jingles that are associated with the olfactory dimensions of a product, e.g., odor pleasantness (Velasco *et al.*, 2014), or the *gestalt* flavor of wine (North, 2012; Spence and Wang, 2015a, b). To be able to broadly implement sound–odor congruence in marketing, the problem of what dimensions of sounds are matched to what dimensions of odors needs to be solved.

One form of congruence that may be important and easy to implement in advertisements is congruence in arousal. Berlyne (1960) considers arousal as "a motivational state pertaining to the level of alertness or activation of an individual, ranging on a continuum from extreme drowsiness to extreme wakefulness" (as cited in Laviolette *et al.*, 2012, p. 728). Arousal is thought to be an autonomic response that is not restricted to a particular modality (Bensafi *et al.*, 2002; Sievers *et al.*, 2017). For example, variations in arousal levels for visual stimuli, sounds, and odors correlate with levels of skin conductance (Bensafi *et al.*, 2002; Khalfa *et al.*, 2002; Lang *et al.*, 1993). The fact that arousal can be evoked via different senses makes it suitable to use in crossmodal correspondence research and in advertisements. Previous studies have successfully manipulated arousal levels of odor and sound in marketing contexts. In cases of sound–odor congruence in arousal, environmental evaluation, approach buying behavior, and overall satisfaction with the shopping experience was enhanced (Mattila and Wirtz, 2001). These results were thought to be due to a positive reaction to a 'coherent environment'. Similarly, sound–odor arousal congruence can result in enhanced pleasure levels, approach behavior, and enhanced overall satisfaction with the shopping experience (Morrison *et al.*, 2011). These studies imply that sound–odor congruence in terms of arousal may be relevant for marketing. However, the generalizability of these studies is limited because only one or two odors and sounds were used. In the present study therefore, we aim to investigate congruence in terms of arousal with a variety of sounds and odors. Despite the suggested importance of congruence in marketing, few studies have investigated its role in advertising. We hypothesize that people's attitudes towards an advertisement, the advertised product, and its brand, will be more positive in cases of sound–odor congruence than in cases of incongruence.

Congruence in an advertisement might also affect product memory, which is important in marketing. Brand memory can have a great impact on consumer choice: If customers remember a product, they are more likely to buy it and talk about it (Hoyer and Brown, 1990). Previous studies suggest congruence in crossmodal correspondences does affect memory. For example, picture–sound congruence can facilitate picture-name retrieval (Chen and Spence, 2010). Similarly, fragrances were better remembered (recognition) when gender of the fragrance (marketed for males versus females) matched the grammatical gender of the fragrance description (Speed and Majid, 2019). However, one study found that congruence of product and scent did not enhance brand recall and recognition (Morrin and Ratneshwar, 2003). Therefore, the extent to which crossmodal congruence affects product memory is not clear.

Product brands can be distinguished between national and private (or store) brands. The market share of private brands has grown in the past decade (Sayman, 2002; Karray and Martin-Herrán, 2009) and private brands are often regarded as competitors of national brands. For crossmodal marketing, the distinction between private and national brands is useful since consumers perceive both types of brands differently (Batra and Sinha, 2000; De Wulf *et al.*, 2005), possibly with different effects on memory. To operationalize this distinction in the present study, we assess familiar and unfamiliar brands (Feetham and Gendall, 2013). We will explore the effect of sound–odor arousal congruence on memory of familiar and unfamiliar brands. It is possible that attitudes towards familiar brands are stronger and more difficult to change (Kent and Allen, 1994; Morrin and Ratneshwar, 2003). We would therefore expect congruence to be more effective for unfamiliar brands than familiar brands (Morrin and Ratneshwar, 2000).

In short, this study investigates whether sound–odor congruence in terms of arousal increases perceived liking of an advertisement, the corresponding brand and product, and whether it enhances brand recognition for familiar versus unfamiliar brands. To test these predictions, we asked participants to smell odors and then view product advertisements paired with a jingle. Afterwards, they rated the advertisement, brand and product, and their memory was tested.

2. Method

2.1. Participants

Fifty-two participants took part in the experiment (31 female, mean age 28, age range 18–68). All participants were Dutch, except for one, who had lived in the Netherlands for over three years, and spoke Dutch at a near-native level. Participants were recruited via social media and word-of-mouth, and participated for free or received course credit for their participation (in case of university students). Based on medium-to-large effects sizes reported elsewhere for similar phenomena (Chen and Spence, 2010; Morrin and Ratneshwar, 2000), a power analysis using G*Power with a medium-to-large effect size ($f^2 = 0.30$) and power of 0.90 indicated a required sample size of 46.

2.2. Materials

Three types of stimuli were manipulated: sounds, odors, and brands. Initially, we selected 24 odors and sounds. The sounds were taken from freesound.org and the odors were either essential oils, like orange oil, or real pieces of fruit, vegetable, plant or tea, e.g., grapefruit. The initial set of odors and sounds

Table 1.

Overall means and standard deviations of the high-arousal and low-arousal odors and sounds of the pre-test ($n = 10$) (1 = low arousal, 9 = high arousal)

	High Arousal	Low Arousal
Odors	6.28 (0.32)	4.29 (0.48)
Sounds	6.79 (0.29)	3.56 (0.37)

was selected to represent a broad range of arousal, based on experimenter intuition. To empirically validate odor and sound arousal, a pre-test was conducted. Based on a pre-test with 10 people, eight odors and sounds rated lowest on arousal and eight odors and sounds rated highest on arousal were selected (measured on SAM-scales; Morris, 1995; see Table 1). Twenty-four products, each with a familiar and unfamiliar brand, were pretested on familiarity. The 16 products (32 brands) from which the familiar and unfamiliar brand varied the most were selected for the experiment. Only brands familiar to Dutch participants were used, and pre-tests were conducted with Dutch participants to ensure that any differences in ratings of the sounds, products, and odors that may arise from cultural factors did not apply to pre-selection of the stimuli.

Sound files were cropped to a duration of 3000 ms using Audacity (see Note 1), so that duration was equated across conditions. This duration was deemed sufficient for participants to perceive the sound and the product image. The high-arousal sounds were clips of high-tempo music and the low-arousal sounds were clips of low-tempo music.

Odors were presented in 30 ml brown opaque jars (height 4.7 cm, diameter 4 cm) and prepared by either putting a few drops of essential oil or pieces of the product on a cotton pad. The cotton pad was then covered with fiber stuffing so participants could not see the content of the jar. A list of the odors and sounds can be found in Appendix A.

A total of 16 products was used and each product had a familiar and unfamiliar brand, like 'Coca Cola' (familiar) and 'Go' (unfamiliar) with the product 'cola'. The products consisted of food, drinks, and body products (i.e., convenience products), because those were expected to be frequently bought with little conscious effort, and thus may be most susceptible to the influence of odor–sound congruence (McDaniel and Baker, 1977). The products were presented as images scaled to the same dimensions (always 336 pixels wide), displayed on a white background. A list of these products can be found in Appendix B. All sounds and images of the products can be found on: https://github.com/ICroijmans/XModalMarketing.

2.3. Procedure

Testing took place in an odor-controlled lab at Radboud University. Participants were informed that the experiment investigated the effectiveness of advertisements. After this, they signed an informed consent form and the experiment began. To get acquainted with the experiment, a practice trial was included. The experiment was performed using E-prime Version 2.0.

In Phase 1 of the experiment, participants were exposed to an odor until they had indicated they could smell it. They then viewed an image of a product with a familiar or unfamiliar brand (as depicted by the product's label), paired with a sound (i.e., an advertisement with a jingle). Half of the advertisements depicted a familiar brand, and half an unfamiliar brand. The order of advertisement presentation was randomized using E-prime. After exposure to the 'advertisement' (product image and sound; 3000 ms), participants rated their attitude towards the advertisement, brand, and product. To measure attitude towards the advertisement, brand, and product, two consumer attitude scales were used based on Voss *et al.* (2003). We used two seven-point semantic differential scales that reflect the 'hedonic' dimension of consumer attitude, ranging from 'boring' (1) to 'exciting' (7), and 'unpleasant' (1) to 'pleasant' (7). The internal consistency of the two items for all three variables was adequate to good: $\alpha = 0.86$, $\alpha = 0.88$, and $\alpha = 0.79$, suggesting reasonable reliability and justifying the use of the mean of both statements in the main analyses. Participants also rated sound and odor pleasantness on a seven-point scale, from 'unpleasant' (1) to 'pleasant' (7). Pleasantness ratings are used as covariates in the analysis. After all stimuli were evaluated, the participants answered questions about whether they listened to upbeat and/or slow music in their spare time and whether they found those types of music relaxing.

In Phase 2 of the experiment, participants' memory for the advertised products was tested. Participants were presented with images of the products from Phase 1, plus 16 new images they had not seen before. The new images were of the same products but with a different brand (e.g., if Coca Cola was seen in Phase 1, the 'new' image in Phase 2 was Go cola). Participants had to respond which brands they had seen before and which were new by clicking on a box on the screen with the word 'new' or 'old' (using a mouse or laptop touchpad). Brands were presented in a random order via E-prime. The entire experiment took around 25 minutes.

Figure 1 depicts the entire procedure. The experiment was approved by the local Ethics Assessment Committee of the Centre for Language Studies, Radboud University, Nijmegen, the Netherlands.

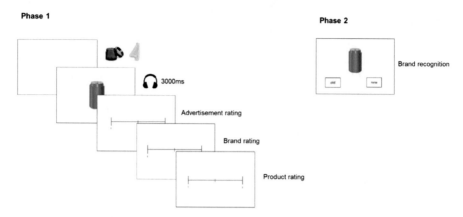

Figure 1. Experimental Procedure. Phase 1: Participants smelled an odor and then were presented with an image of a product and a sound clip. They then rated their attitude towards the advertisement, brand, and product. Phase 2: Participants were presented with images of products from Phase 1 and an equal number of new products and had to decide which they had seen before.

3. Results

3.1. Attitude Towards the Advertisement, Brand and Product

Separate linear mixed effects models (LME) were conducted using the lme4 package (Bates *et al.*, 2015) in R (R Core Team, 2013) on the three measures of attitude (advertisement, brand, and product) with participants and items as random effects, odor arousal, sound arousal and familiarity as fixed factors and odor pleasantness ratings and sound pleasantness ratings as covariates. For all analyses non-standardized betas are reported.

There was no interaction between sound and odor arousal on attitude towards the advertisement, brand or product. Therefore, there was no sound–odor congruency effect on attitude. However, there was a significant main effect of sound arousal ($t = 5.90$, $p < 0.001$, $\beta = 0.81$) on attitude towards the advertisement. Advertisements with high-arousing sounds ($M = 4.39$, SD = 1.15) were more positively evaluated than those with low-arousing sounds ($M = 3.68$, SD = 1.18). In addition, there was a significant main effect of familiarity ($t = 5.55$, $p < 0.001$, $\beta = 0.75$) on brand attitude. The attitude towards familiar brands was more positive ($M = 4.49$, SD = 1.19) than towards unfamiliar brands ($M = 3.63$, SD = 1.13). Lastly, there was a significant main effect of familiarity on product attitude ($t = 2.86$, $p = 0.004$, $\beta = 0.41$). Product attitude was more positive in cases of familiar brands ($M = 4.44$, SD = 1.18) compared with unfamiliar brands ($M = 3.98$, SD = 1.26). There were no other significant effects (Table 2).

Table 2.

Means and standard deviations of attitude towards the advertisement across conditions (1 = boring/unpleasant, 7 = exciting/pleasant)

Type of Attitude		Familiar brand		Unfamiliar brand	
		High Arousal Sound	Low Arousal Sound	High Arousal Sound	Low Arousal Sound
Advertisement	High Arousal Odor	4.45 (1.19)	3.72 (1.19)	4.12 (1.20)	3.54 (1.23)
	Low Arousal Odor	4.51 (1.04)	3.86 (1.18)	4.48 (1.12)	3.60 (1.09)
Brand	High Arousal Odor	4.57 (1.24)	4.35 (1.20)	3.73 (1.19)	3.42 (1.00)
	Low Arousal Odor	4.55 (1.16)	4.49 (1.18)	3.88 (1.15)	3.50 (1.11)
Product	High Arousal Odor	4.40 (1.28)	4.43 (1.15)	3.88 (1.18)	3.85 (1.26)
	Low Arousal Odor	4.62 (1.13)	4.30 (1.15)	4.24 (1.30)	3.94 (1.27)

3.2. Product Recognition

The same LME analyses as above were repeated for recognition accuracy and reaction times. For recognition accuracy, a log-linear model was used. Recognition accuracy was measured as the number of correct responses. Reaction times were log-transformed, but for interpretation purposes, raw reaction times and standard deviations are reported. Reaction times were recorded from the onset of the product image to the time when a participant made their response.

Since none of the effects for recognition accuracy were significant, only the analysis for reaction times is reported (see Table 3). There was a main effect of sound arousal ($t = 2.26$, $p = 0.024$, $\beta = 0.05$), odor arousal ($t = 2.12$, $p = 0.034$, $\beta = 0.05$) and familiarity ($t = 3.93$, $p < 0.001$, $\beta = 0.09$). Products were recognized faster with low-arousal sounds ($M = 1673$ ms, SD $= 870$) compared to high-arousal sounds ($M = 1687$ ms, SD $= 942$), and with low-arousal odors ($M = 1656$ ms, SD $= 834$) compared to high-arousal odors ($M = 1704$ ms, SD $= 973$). Familiar brands were recognized more quickly ($M = 1602$ ms, SD $= 868$) than unfamiliar brands ($M = 1758$ ms, SD $= 937$).

Table 3.

Means and standard deviations of reaction times (ms) across conditions

	Familiar brand		Unfamiliar brand	
	High Arousal Sound	Low Arousal Sound	High Arousal Sound	Low Arousal Sound
High Arousal Odor	1437 (587)	1743 (1110)	1906 (1181)	1741 (882)
Low Arousal Odor	1654 (999)	1581 (663)	1766 (868)	1629 (779)

There was also a significant interaction between sound arousal and familiarity ($t = 2.34$, $p = 0.019$, $\beta = 0.08$). A follow-up analysis showed that when sound arousal was high, familiar brands ($M = 1544$ ms, SD $= 821$) were recognized more quickly than unfamiliar brands ($M = 1837$ ms, SD $= 1036$) ($t = 3.90$, $p < 0.001$, $\beta = 0.09$). There was no difference in familiarity for low-arousal sounds. There was a significant interaction between odor arousal and familiarity ($t = 2.35$, $p = 0.019$, $\beta = 0.09$). For high-arousal odors, familiar brands were recognized more quickly ($M = 1586$ ms, SD $= 893$) than unfamiliar brands ($M = 1820$ ms, SD $= 1037$) ($t = 3.83$, $p < 0.001$, $\beta = 0.09$). No significant difference in familiarity was found for low-arousal odors. Overall, brand recognition was facilitated for familiar brands when arousal was high (in both odor and sound).

Finally, there was also a significant interaction between odor arousal and sound arousal ($t = 2.20$, $p = 0.028$, $\beta = 0.07$). A follow-up analysis revealed that when sound arousal was high, brands were more quickly recognized with high-arousal odors ($M = 1665$ ms, SD $= 951$) than low-arousal odors ($M = 1709$ ms, SD $= 936$) ($t = 2.46$, $p = 0.014$, $\beta = 0.06$). In addition, when odor arousal was high, brands were more quickly recognized in case of high-arousal sounds ($M = 1665$ ms, SD $= 951$) than in case of low-arousal sounds ($M = 1742$ ms, SD $= 996$) ($t = 2.19$, $p = 0.029$, $\beta = 0.05$). No further significant effects appeared when odor arousal or sound arousal was low. Overall, sound–odor congruence positively affected product memory, but only in case of high-arousal congruence.

4. General Discussion

This study assessed whether crossmodal congruence in terms of arousal between odors and sounds affected evaluation of advertisements and the advertised brands and products, and memory for the products, for familiar and unfamiliar brands. We found that sound–odor congruence affected memory for advertised products. When odor and sound were congruent in terms of high arousal, product recognition was faster than in the case of incongruence.

In contrast to previous studies (Mattila and Wirtz, 2001; Morrison *et al.*, 2011) we found no effect of sound–odor congruence on attitudes towards the advertisement, brand or product. This could be due to the timing of our stimuli, since crossmodal correspondences are most likely to be experienced when the stimuli are presented around the same time (Spence, 2011). Although the odors and sounds were presented quickly after one another, they were not presented simultaneously, which could reduce the likelihood of a congruence effect (e.g., Meredith *et al.*, 1987). A follow-up study using an olfactometer to synchronize presentation of the odor and sound stimuli may yield different results.

High-arousal sounds led to more positive evaluations of the advertisement overall than low-arousal sounds. This could be because it is more common in advertisements to use upbeat music than slow tempo music, illustrating the familiarity or mere-exposure principle (Zajonc, 1968). In line with this, previous experiments showed that arousing advertisements produce more positive attitudes towards advertised brands and lead to more purchase intentions (Yoon *et al.*, 1998). The lack of effects of sound arousal on brand and product attitude might be explained by the fact that the sound fragment was perceived as part of the advertisement, making the link between sound and advertisement more salient than the link between sound and brand, or sound and product.

Contrary to sound, the level of odor arousal did not affect the evaluation of the advertisement (brand and product) evaluations, suggesting the effect of sound arousal was not due to a general arousal effect. Brand communication typically occurs in only two sensory channels: vision and hearing (Lindstrom, 2005), therefore, people may not be used to linking what they smell to what they hear and see in an advertisement. Furthermore, the odor was presented before the advertisement and therefore may not have been fully integrated with it (e.g., Meredith *et al.*, 1987), whereas the sound may have been bound together with the advert due to their temporal synchrony. Another important factor may be the difference between high- and low-arousal stimuli for odors and sounds. According to the mean ratings, high- and low-arousal sounds differed more in arousal than high- and low-arousal odors did (see Table 1). It is possible then that high- and low-arousal odors did not differ enough to influence perception of the advert.

People's attitude towards the advertisement, brand and product was more positive for familiar than unfamiliar brands (indicating that familiarity was successfully manipulated). Again, this might be explained by the mere-exposure principle (Zajonc, 1968): as people's exposure to stimuli increases, their attitude towards those stimuli becomes more positive. On the other hand, people's attitudes towards unfamiliar brands might be neutral, as they have not had sufficient time to form an impression of the brand. This could explain why the evaluation of unfamiliar brands ($M = 3.85$) is closer to the middle of the seven-point Likert-scale and consequently lower than that of familiar brands ($M = 4.35$). This underscores the fact that new brands need to work to achieve positive consumer evaluations.

Although we did not find any differences in recognition accuracy, there were differences in recognition reaction times, which is likely to be a more sensitive measure to detect potential effects of crossmodal correspondences on memory. A sound–odor congruence effect was found on reaction time for high-arousal stimuli: there was an advantage of congruence over incongruence when both sound arousal and odor arousal were high. This could be explained by the difference between the desired or expected arousal level and the actual arousal level (Wirtz *et al.*, 2007). When participants smelled a high-arousal odor, it led to the formation of an expected arousal level. When the following sound fragment turned out to be low-arousal, under-stimulation created a discrepancy between level of arousal evoked by the sound and the expected/desired arousal level based on the odor. No such discrepancy occurred when the following sound was high-arousal. Likewise, sequential exposure to a low-arousal odor and high-arousal sound might have led to over-stimulation. High-arousal congruence could have led to the realization of the desired arousal level, potentially enabling the creation of a stronger memory trace. Another possibility is that congruence leads to feelings of familiarity (e.g., Zajonc, 1968), or the forming of a coherent sensory gestalt (Stach, 2015), or memory 'chunk' (Mathy and Feldman, 2012; Miller, 1956), both of which should facilitate recognition (Mandler, 1980; Miller, 1956). However, according to these arguments, the same effect should have occurred for low-arousal congruence, which was not the case. We note, however, that the difference in arousal ratings between low-arousal sounds and low-arousal odors was bigger than the difference between high-arousal sounds and high-arousal odors. Perhaps low-arousal sounds and odors were not truly 'congruent' in the same way the high-arousal stimuli were.

There are other possible explanations for why congruence in terms of high arousal only facilitated brand recognition. In general, high-arousal stimuli are remembered better than low-arousal stimuli (e.g., Bradley *et al.*, 1992), and the combination of high-arousal odor and high-arousal sound in the present study may have had an additive effect. The finding is also in line with research elsewhere that found that background music affected food perception for emotional food (milk chocolate) but not neutral food (red pepper; Fiegel *et al.*, 2014). Emotion might be a moderator in such a way that congruence only emerges when the stimuli are not neutral. Another possibility is that adverts that have high odor arousal and high sound arousal are more interesting (Baker *et al.*, 1992). If so, participants may have engaged with these adverts more, resulting in stronger encoding.

In addition to congruence effects, familiar brands were recognized more quickly than unfamiliar brands. This effect only occurred when sound arousal or odor arousal was high. High sound or odor arousal appeared to enhance the difference between familiar and unfamiliar brands, but both stimuli did not have a positive effect on memory alone. It is possible that the effect of arousal on memory depends on the difficulty of the memory task: in easy tasks (e.g., remembering familiar brands) arousal may aid memory, whereas for memory of more difficult stimuli (e.g., unfamiliar brands) it could interfere.

Our results have important implications for marketing. High-arousal sound–odor congruence led to faster product recognition times, which could be relevant for consumers' product memory in novel ways. For example, a relatively new innovation, the smelling screen (Aravinda and Krishnaiah, 2013), could be used to present adverts in shops using both sound and smell. The screen could distribute a high-arousal odor in store and simultaneously play high-arousal music, helping people to remember the product and recognize it better when they see it again. A similar approach could also be used for aroma billboards and sounds. Additionally, cinemas pose a relatively closed-off environment that may be suitable for smell–sound marketing applications, with the atmospheric composition easy to monitor, and relatively easy to control (cf. Williams *et al*, 2016). Although presenting smells in cinemas still needs to overcome a range of problems with delivery of the smell, and possible physiological drawbacks of it (see Spence, Obrist, Velasco & Ranasinghe, 2017, for a review), presenting a single smell along with a trailer or commercial in cinemas may be another possible avenue where smell–sound marketing could be applied.

This study has demonstrated the potential marketing effects of sound and odor separately as well as combined. More insight is needed into how arousal can be used to stimulate product memory, because it may work differently depending on the familiarity of the brand. In addition, it is not clear why we found a congruence effect on memory with high-arousal stimuli only. It could be possible that low-arousal congruence leads to different effects for other types of products (for example, for purchases that require more rumination). Future studies could also investigate whether similar effects occur in different marketing settings. In particular, creating marketing contexts that are more ecologically valid would be important (for example, by using virtual supermarkets). This will increase marketers' knowledge of customer behavior and help them to design novel and unexpected strategies to tackle the growing need to stand out in an increasingly competitive environment. Overall, there is vast potential for the implementation of sound and odor in marketing innovations.

Acknowledgements

IC and LJS were supported by The Netherlands Organization for Scientific Research: NWO VICI grant "Human olfaction at the intersection of language, culture and biology", project number 277-70-011, awarded to Asifa Majid.

Note

1. https://www.audacityteam.org/

References

Aravinda, C. and Krishnaiah, R. V. (2013). Smell-o-vision — the future digital display device, *Int J. Comput. Sci. Mobile Comput.* **2**, 227–234.

Auvray, M. and Spence, C. (2008). The multisensory perception of flavor, *Consc. Cogn.* **17**, 1016–1031.

Baker, J., Levy, M. and Grewal, D. (1992). An experimental approach to making retail store environmental decisions, *J. Retail* **68**, 445–460.

Bates, D., Maechler, M., Bolker, B. and Walker, S. (2015). Fitting linear mixed-effects models using lme4, *J. Stat. Softw.* **67**, 1–48.

Batra, R. and Sinha, I. (2000). Consumer-level factors moderating the success of private label brands, *J. Retail.* **76**, 175–191.

Belkin, K., Martin, R., Kemp, S. E. and Gilbert, A. N. (1997). Auditory pitch as a perceptual analogue to odor quality, *Psychol. Sci.* **8**, 340–342.

Bensafi, M., Rouby, C., Farget, V., Bertrand, B., Vigouroux, M. and Holley, A. (2002). Autonomic nervous system responses to odours: the role of pleasantness and arousal, *Chem. Senses* **27**, 703–709.

Berlyne, D. (1960). *Conflict, Arousal, and Curiosity*. McGraw-Hill, New York, NY, USA.

Biswas, D. and Szocs, C. (2019). The smell of healthy choices: cross-modal sensory compensation effects of ambient scent on food purchases, *J. Mark. Res.* **56**, 123–141.

Biswas, D., Lund, K. and Szocs, C. (2019). Sounds like a healthy retail atmospheric strategy: effects of ambient music and background noise on food sales, *J. Acad. Mark. Sci.* **47**, 37–55.

Bradford, K. D. and Desrochers, D. M. (2009). The use of scents to influence consumers: the sense of using scents to make cents, *J. Bus. Ethics* **90**, 141–153.

Bradley, M. M., Greenwald, M. K., Petry, M. C. and Lang, P. J. (1992). Remembering pictures: pleasure and arousal in memory, *J. Exp. Psychol. Learn. Mem. Cogn.* **18**, 379–390.

Chen, Y.-C. and Spence, C. (2010). When hearing the bark helps to identify the dog: semantically-congruent sounds modulate the identification of masked pictures, *Cognition* **114**, 389–404.

Cheng, J., Sun, A. and Zeng, D. (2010). Information overload and viral marketing: countermeasures and strategies, in: *Third International Conference on Social Computing, Behavioral Modeling and Prediction, SPB 2010*, pp. 108–117. Advances in Social Computing, Bethesda, MD, USA.

Crisinel, A.-S., Cosser, S., King, S., Jones, R., Petrie, J. and Spence, C. (2012). A bittersweet symphony: systematically modulating the taste of food by changing the sonic properties of the soundtrack playing in the background, *Food Qual. Pref.* **24**, 201–204.

Crisinel, A.-S., Jacquier, C., Deroy, O. and Spence, C. (2013). Composing with cross-modal correspondences: music and odors in concert, *Chemosens. Percept.* **6**, 45–52.

Davies, B. J., Kooijman, D. and Ward, P. (2003). The sweet smell of success: olfaction in retailing, *J. Mark. Manag.* **19**, 611–627.

de Lange, M. A., Debets, L. W., Ruitenburg, K. and Holland, R. W. (2012). Making less of a mess: scent exposure as a tool for behavioral change, *Soc. Influ.* **7**, 90–97.

De Wijk, R. A., Smeets, P. A., Polet, I. A., Holthuysen, N. T., Zoon, J. and Vingerhoeds, M. H. (2018). Aroma effects on food choice task behavior and brain responses to bakery food product cues, *Food Qual. Pref.* **68**, 304–314.

De Wulf, K., Odekerken-Schröder, G., Goedertier, F. and Van Ossel, G. (2005). Consumer perceptions of store brands *versus* national brands, *J. Consum. Mark.* **22**, 223–232.

Deroy, O., Crisinel, A.-S. and Spence, C. (2013). Crossmodal correspondences between odors and contingent features: odors, musical notes, and geometrical shapes, *Psychon. Bull. Rev.* **20**, 878–896.

Dyer, G. (1982). *Advertising as Communication*. Routledge, London, UK.

Feetham, P. and Gendall, P. (2013). The positioning of premium private label brands, *Mark. Soc. Res.* **21**, 28–37.

Fiegel, A., Meullenet, J.-F., Harrington, R. J., Humble, R. and Seo, H.-S. (2014). Background music genre can modulate flavor pleasantness and overall impression of food stimuli, *Appetite* **76**, 144–152.

Fransen, M. L., Verlegh, P. W. J., Kirmani, A. and Smit, E. G. (2015). A typology of consumer strategies for resisting advertising, and a review of mechanisms for countering them, *Int. J. Advert.* **34**, 6–16.

Gabrielsson, A. (2001). Emotions in strong experiences with music, in: *Music and Emotion: Theory and Research*, P. N. Juslin and J. A. Sloboda (Eds), pp. 431–449. Oxford University Press, Oxford, UK.

Hagtvedt, H. and Brasel, S. A. (2016). Cross-modal communication: sound frequency influences consumer responses to color lightness, *J. Mark. Res.* **53**, 551–562.

Holland, R. W., Hendriks, M. and Aarts, H. (2005). Smells like clean spirit: nonconscious effects of scent on cognition and behavior, *Psychol. Sci.* **16**, 689–693.

Hoyer, W. D. and Brown, S. P. (1990). Effects of brand awareness on choice for a common, repeat-purchase product, *J. Consum. Res.* **17**, 141–148.

Karray, S. and Martín-Herrán, G. (2009). A dynamic model for advertising and pricing competition between national and store brands, *Eur. J. Oper. Res.* **193**, 451–467.

Kellaris, J. J. and Kent, R. J. (1993). An exploratory investigation of responses elicited by music varying in tempo, tonality, and texture, *J. Consum. Psychol.* **2**, 381–401.

Kent, R. J. and Allen, C. T. (1994). Competitive interference effects in consumer memory for advertising: the role of brand familiarity, *J. Market.* **58**, 97–105.

Khalfa, S., Isabelle, P., Jean-Pierre, B. and Manon, R. (2002). Event-related skin conductance responses to musical emotions in humans, *Neurosci. Lett.* **328**, 145–149.

Krishna, A., Elder, R. S. and Caldara, C. (2010). Feminine to smell but masculine to touch? Multisensory congruence and its effect on aesthetic experience, *J. Consum. Psychol.* **20**, 410–418.

Lang, P. J., Greenwald, M. K., Bradley, M. M. and Hamm, A. O. (1993). Looking at pictures: affective, facial, visceral, and behavioral reactions, *Psychophysiology* **30**, 261–273.

Laviolette, E. M., Radu Lefebvre, M. and Brunel, O. (2012). The impact of story bound entrepreneurial role models on self-efficacy and entrepreneurial intention, *Int. J. Entrep. Behavi. Res.* **18**, 720–742.

Liljenquist, K., Zhong, C.-B. and Galinsky, A. D. (2010). The smell of virtue: clean scents promote reciprocity and charity, *Psychol. Sci.* **21**, 381–383.

Lindstrom, M. (2005). Broad sensory branding, *Prod. Brand Manag.* **14**, 84–87.

Mandler, G. (1980). Recognizing: the judgment of previous occurrence, *Psychol. Rev.* **87**, 252–271.

Mandler, G. (1982). The structure of value: accounting for taste, in: *Affect and Cognition: the 17th Annual Carnegie Symposium*, M. S. Clark and S. T. Fiske (Eds), pp. 3–36. Lawrence Erlbaum Associates, Hillsdale, NJ, USA.

Mathy, F. and Feldman, J. (2012). What's magic about magic numbers? Chunking and data compression in short-term memory, *Cognition* **122**, 346–362.

Mattila, A. S. and Wirtz, J. (2001). Congruency of scent and music as a driver of in-store evaluations and behavior, *J. Retail.* **77**, 273–289.

McDaniel, C. and Baker, R. C. (1977). Convenience food packaging and the perception of product quality, *J. Market.* **41**, 57–58.

Meredith, M. A., Nemitz, J. W. and Stein, B. E. (1987). Determinants of multisensory integration in superior colliculus neurons. I. Temporal factors, *J. Neurosci.* **7**, 3215–3229.

Meyers-Levy, J. and Tybout, A. M. (1989). Schema congruity as a basis for product evaluation, *J. Consum. Res.* **16**, 39–54.

Mikutta, C. A., Schwab, S., Niederhauser, S., Wuermle, O., Strik, W. and Altorfer, A. (2013). Music, perceived arousal, and intensity: psychophysiological reactions to Chopin's "Tristesse", *Psychophysiology* **50**, 909–919.

Miller, G. A. (1956). The magical number seven, plus or minus two: some limits on our capacity for processing information, *Psychol. Rev.* **63**, 81–97.

Mitchell, D. J., Kahn, B. E. and Knasko, S. C. (1995). There's something in the air: effects of congruent or incongruent ambient odor on consumer decision making, *J. Consum. Res.* **22**, 229–238.

Moeran, B. (2007). Marketing scents and the anthropology of smell, *Soc. Anthrop.* **15**, 153–168.

Morrin, M. and Ratneshwar, S. (2000). The impact of ambient scent on evaluation, attention, and memory for familiar and unfamiliar brands, *J. Bus. Res.* **49**(2), 157–165.

Morrin, M. and Ratneshwar, S. (2003). Does it make sense to use scents to enhance brand memory? *J. Mark. Res.* **40**, 10–25.

Morris, J. D. (1995). Observations: SAM: the self-assessment manikin: an efficient crosscultural measurement of emotional response, *J. Advert. Res.* **35**, 63–68.

Morris, J. D. and Boone, M. A. (1998). The effects of music on emotional response, brand attitude, and purchase intent in an emotional advertising condition, *Adv. Consum. Res.* **25**, 518–526.

Morrison, M., Gan, S., Dubelaar, C. and Oppewal, H. (2011). In-store music and aroma influences on shopper behavior and satisfaction, *J. Bus. Res.* **64**, 558–564.

North, A. C., Hargreaves, D. J. and McKendrick, J. (1999). The influence on in-store music on wine selections, *J. Appl. Psychol.* **84**, 271–276.

North, A. C. (2012). The effect of background music on the taste of wine, *Br. J. Psychol.* **103**, 293–301.

Petruzzellis, L., Chebat, J.-C. and Palumbo, A. (2014). "Hey dee-jay let's play that song and keep me shopping all day long": the effect of famous background music on consumer shopping behavior, *J. Mark. Dev. Compet.* **8**, 38–49.

Piqueras-Fiszman, B. and Spence, C. (2015). Sensory expectations based on product-extrinsic food cues: an interdisciplinary review of the empirical evidence and theoretical accounts, *Food Qual. Prefer.* **40**, 165–179.

R Core Team (2013). *R: a Language and Environment for Statistical Computing.* R Foundation for Statistical Computing, Vienna, Austria.

Reinoso-Carvalho, F., Van Ee, R., Rychtarikova, M., Touhafi, A., Steenhaut, K., Persoone, D., Spence, C. and Leman, M. (2015). Does music influence the multisensory tasting experience? *J. Sens. Stud.* **30**, 404–412.

Rickard, N. S. (2004). Intense emotional responses to music: a test of the physiological arousal hypothesis, *Psychol. Music* **32**, 371–388.

Sayman, S., Hoch, S. J. and Raju, J. S. (2002). Positioning of store brands, *Market. Sci.* **21**, 378–397.

Seo, H. S. and Hummel, T. (2011). Auditory–olfactory integration: congruent or pleasant sounds amplify odor pleasantness, *Chem. Senses* **36**, 301–309.

Sievers, B., Lee, C., Haslett, W. and Wheatley, T. (2017). A multi-sensory code for arousal. https://psyarxiv.com/wucs4/download?format=pdf.

Simner, J., Cuskley, C. and Kirby, S. (2010). What sound does that taste? Cross-modal mappings across gustation and audition, *Perception* **39**(4), 553–569.

Small, D. M. and Prescott, J. (2005). Odor/taste integration and the perception of flavor, *Exp. Brain Res.* **166**, 345–357.

Speed, L. J. and Majid, A. (2019). Linguistic features of fragrances: the role of grammatical gender and gender associations. *Atten. Percept. Psychophys.* http://hdl.handle.net/21.11116/0000-0003-4379-A.

Spence, C. (2011). Crossmodal correspondences: a tutorial review, *Atten. Percept. Psychophys.* **73**, 971–995.

Spence, C. and Parise, C. V. (2012). The cognitive neuroscience of crossmodal correspondences, *i-Perception* **3**, 410–412.

Spence, C. (2019). Multisensory experiential wine marketing, *Food Qual. Pref.* **71**, 106–116.

Spence, C., Obrist, M., Velasco, C. and Ranasinghe, N. (2017). Digitizing the chemical senses: Possibilities & pitfalls, *International Journal of Human-Computer Studies* **107**, 62–74.

Spence, C. and Wang, Q. J. (2015a). Wine and music (I): on the crossmodal matching of wine and music, *Flavour* **4**, 34. DOI:10.1186/s13411-015-0045-x.

Spence, C. and Wang, Q. J. (2015b). Wine and music (II): can you taste the music? Modulating the experience of wine through music and sound, *Flavour* **4**, 33. DOI:10.1186/s13411-015-0043-z.

Stach, J. (2015). A conceptual framework for the assessment of brand congruent sensory modalities, *J. Brand Manag.* **22**, 673–694.

Treisman, A. (1996). The binding problem, *Curr. Opin. Neurobiol.* **6**, 171–178.

Treisman, A. (1998). Feature binding, attention and object perception, *Phil. Trans. R. Soc. Lond. B Biol. Sci.* **353**, 1295–1306.

Velasco, C., Balboa, D., Marmolejo-Ramos, F. and Spence, C. (2014). Crossmodal effect of music and odor pleasantness on olfactory quality perception, *Front. Psychol.* **5**, 1352. DOI:10.3389/fpsyg.2014.01352.

Velasco, C. and Spence, C. (2019). The multisensory analysis of product packaging framework, in: *Multisensory Packaging*, C. Velasco and C. Spence (Eds), pp. 191–223. Palgrave Macmillan, Cham, Switzerland.

Voss, K. E., Spangenberg, E. R. and Grohmann, B. (2003). Measuring the hedonic and utilitarian dimensions of consumer attitude, *J. Market. Res.* **40**, 310–320.

Wang, Q. and Spence, C. (2016). 'Striking a sour note': assessing the influence of consonant and dissonant music on taste perception, *Multisens. Res.* **29**, 195–208.

Williams, J., Stönner, C., Wicker, J., Krauter, N., Derstroff, B., Bourtsoukidis, E., Klüpfel, T. and Kramer, S. (2016). Cinema audiences reproducibly vary the chemical composition of air during films, by broadcasting scene specific emissions on breath, *Sci. Rep.* **6**, 25464. DOI:10.1038/srep25464.

Wirtz, J., Mattila, A. S. and Tan, R. L. P. (2007). The role of arousal congruency in influencing consumers' satisfaction evaluations and in-store behaviors, *Int. J. Serv. Ind. Manag.* **18**, 6–24.

Yeshurun, Y. and Sobel, N. (2010). An odor is not worth a thousand words: from multidimensional odors to unidimensional odor objects, *Annu. Rev. Psychol.* **61**, 219–241.

Yoon, K., Bolls, P. and Lang, A. (1998). The effects of arousal on liking and believability of commercials, *J. Mark. Commun.* **4**, 101–114.

Zajonc, R. B. (1968). Attitudinal effects of mere exposure, *J. Pers. Soc. Psychol.* **9**, 1–27.

Zellner, D., Geller, T., Lyons, S., Pyper, A. and Riaz, K. (2017). Ethnic congruence of music and food affects food selection but not liking, *Food Qual. Pref.* **56**, 126–129.

Appendix A.

The odors that were selected for the experiment are in bold. Sound fragments can be found on https://github.com/ICroijmans/XModalMarketing

Odor	Pleasantness *M*	Pleasantness SD	Arousal *M*	Arousal SD
Low arousal				
Cucumber	4.8	1.62	3.7	2.06
Leather	5.1	1.66	3.7	1.95
Green tea	4	1.7	4	1.56
Grapefruit	5.6	1.08	4.1	1.66
Milk	3.1	1.6	4.5	2.42
Coconut	5.9	1.85	4.6	2.07
Soap	7.2	0.63	4.7	2.5
Jasmine	6.1	0.74	5	1.16
High arousal				
Garlic	2.7	1.06	5.9	2.28
Gasoline	4.3	2.36	6	2.11
Mint	6.2	2.4	6	2.16
Vanilla	5.7	1.64	6.1	1.6
Coffee	6.1	2.42	6.3	2
Orange	7.2	1.23	6.6	0.84
Valerian Root	2.4	1.43	6.6	1.08
Chamomile	2.3	1.16	6.7	2.21

Appendix B.
Products and corresponding familiar and unfamiliar brands used in the experiment

Product		Mean Familiar	SD Familiar		Mean Unfamiliar	SD Unfamiliar
Water	*Spa*	6.9	0.32	*Theoni*	*1.5*	1.58
Spinach	*Iglo*	6.4	0.84	*AllSeasons*	1.5	0.71
Coke	*Coca-Cola*	7	0	*Go Cola*	1.1	0.32
Orange Juice	*Appelsientje*	6.8	0.63	*Goldhorn*	*1.1*	0.32
Chocolate Sprinkles	*Venz*	6.2	1.87	*Riox*	1.7	1.49
Peas	*Bonduelle*	6.6	0.7	*Green*	1.4	0.7
Cornflakes	*Kellogg's*	6.4	0.7	*Tilo's*	1.3	0.68
Tomato soup	*Unox*	6.8	0.42	*Calide*	1	0
Tissues	*Kleenex*	6.4	0.84	*Sublimo*	1.1	0.32
Body cream	*Nivea*	6.8	0.42	*Cetaphil*	1	0
Ketchup	*Heinz*	6.9	0.32	*Del Monte*	1.9	1.91
Washing-up liquid	*Dreft*	6.7	0.48	*Easy*	1.6	1.9
Detergent	*Robijn*	6.7	0.48	*Cheer*	*1*	0
Olive oil	*Bertolli*	6.6	0.7	*Vigo*	1.6	1.58
Crisps	*Lays*	6.8	0.63	*Fine Food*	1.1	0.32
Milk	*Campina*	6.8	0.42	*Milsa*	2.1	1.66

Appendix C.
Questionnaire (translated from Dutch to English)

Attitudes

Attitude towards the advertisement
I found the advertisement.

Boring	1	2	3	4	5	6	7	Exciting
Unpleasant	1	2	3	4	5	6	7	Pleasant

Brand attitude
I found the brand of the advertisement.

Boring	1	2	3	4	5	6	7	Exciting
Unpleasant	1	2	3	4	5	6	7	Pleasant

Product attitude
I found the product of the advertisement.

Boring	1	2	3	4	5	6	7	Exciting
Unpleasant	1	2	3	4	5	6	7	Pleasant

Covariates

Odor pleasantness
I found the odor.

Unpleasant	1	2	3	4	5	6	7	Pleasant

Sound pleasantness
I found the sound fragment of the advertisement.

Unpleasant	1	2	3	4	5	6	7	Pleasant

Music preference

In my spare time I often listen to upbeat music

Strongly disagree	1	2	3	4	5	6	7	Strongly agree

In my spare time I often listen to slow temp music

Strongly disagree	1	2	3	4	5	6	7	Strongly agree

Upbeat music relaxes me.

Strongly disagree	1	2	3	4	5	6	7	Strongly agree

Slow tempo music relaxes me.

Strongly disagree	1	2	3	4	5	6	7	Strongly agree

Not Just Another Pint! The Role of Emotion Induced by Music on the Consumer's Tasting Experience

Felipe Reinoso-Carvalho [1,2,*], **Silvana Dakduk** [1], **Johan Wagemans** [2] and **Charles Spence** [3]

[1] School of Management, Universidad de los Andes, Calle 21 # 1-20, Edificio SD, Room SD-940, Bogotá, Colombia

[2] Brain and Cognition, University of Leuven (KU Leuven), Tiensestraat 102 — box 3711, Leuven B-3000, Belgium

[3] Crossmodal Research Laboratory, University of Oxford, Anna Watts Building, Oxford, OX2 6GG, UK

Abstract

We introduce a novel methodology to assess the influence of the emotion induced by listening to music on the consumer's multisensory tasting experience. These crossmodal effects were analyzed when two contrasting music tracks (positive vs negative emotion) were presented to consumers while tasting beer. The results suggest that the emotional reactions triggered by the music influenced specific aspects of the multisensory tasting experience. Participants liked the beer more, and rated it as tasting sweeter, when listening to music associated with positive emotion. The same beer was rated as more bitter, with higher alcohol content, and as having more body, when the participants listened to music associated with negative emotion. Moreover, participants were willing to pay 7–8% more for the beer that was tasted while they listened to positive music. This novel methodology was subsequently replicated with two different styles of beer. These results are discussed along with practical implications concerning the way in which music can add significant value to how a consumer responds to a brand.

Keywords

Beer, flavor, music, crossmodal correspondences, sensory marketing, multisensory

---

[*] To whom correspondence should be addressed. E-mail: f.reinosoc@uniandes.edu.co; f.sound@gmail.com

© KONINKLIJKE BRILL NV, LEIDEN, 2019 | DOI:10.1163/9789004416307_007

1. Introduction

The multisensory nature of tasting has become increasingly clear to researchers in recent years (e.g., Auvray and Spence, 2008; Piqueras-Fiszman and Spence, 2016). As a matter of fact, renowned researchers from the field of marketing and consumer behavior have further enriched the discussion concerning how such novel insights (which are mostly being developed from the emerging cognitive neuroscience research on multisensory perception) can be brought to bear in the context of multisensory marketing and branding (e.g., Hultén, 2011; Krishna, 2012; Spence, 2019a).

A growing body of research is now addressing the question of how what is heard influences the taste and flavor of foods and beverages (e.g., see Corr and Plagnol, 2018, Chapter 7). The research that has been conducted suggests that certain types of sounds and music can — at least under the appropriate conditions — add significant value, and pleasure, to the eating/drinking experience, whereas the wrong music can also impair enjoyment and/or negatively affect the perceived value of such experiences (see Spence, 2017a, for a review). These studies have highlighted multisensory factors influencing the tasting experience and, to date, several different methods have been proposed to assess the effect of what we hear on what we taste (see Knöferle and Spence, 2012, for a review).

On the one hand, a spate of studies has highlighted the influence of the sound of the food itself, considering that this can add significant value (not to mention pleasure) to the consumer's overall multisensory eating/drinking experience (e.g., Spence *et al.*, 2011). On the other hand, the sound and/or noise in those places in which we eat and drink — such as restaurants, airplanes — have been shown to dramatically affect the way in which we rate taste and flavor (Spence, 2012, 2014, 2017b; Spence *et al.*, 2014).

The interactions between what we hear and what we eat/drink are particularly intriguing because it is not immediately obvious how, or even why, what we hear should influence what we taste, in those cases where the inputs from the various senses share nothing in common (as when listening to music or ambient soundscapes while eating/drinking). For instance, researchers have managed to isolate a number of specific sonic and musical parameters (such as pitch and instrumentation) that can be used to modify the tasting experience in predictable ways (e.g., Bronner *et al.*, 2012; Crisinel and Spence, 2009, 2010, 2012; Crisinel *et al.*, 2012; Reinoso Carvalho *et al.*, 2016a). For example, Reinoso Carvalho *et al.* (2016a) demonstrated that people tend to associate sweet tastes with a higher pitch, whereas bitter tastes are usually associated with lower-pitched sounds instead (cf. Crisinel and Spence, 2010; Holt-Hansen, 1968, 1976; Rudmin and Cappelli, 1983).

Based on a growing list of such crossmodal associations, a range of taste/flavor-related music tracks have now been composed by artists, designers, sonic branding agencies, and researchers (e.g., Knoeferle *et al.*, 2015; Mesz *et al.*, 2011; Reinoso Carvalho *et al.*, 2015a, 2016b; Wang and Spence, 2016). Some of these studies have further investigated the influence of this type of music on the way in which consumers judge different aspects of food (e.g., Reinoso Carvalho, 2015a, b; Wang and Spence, 2016), and drink (Reinoso Carvalho *et al.*, 2016b; Spence *et al.*, 2013; Wang and Spence, 2015a, b). For example, when it comes to alcoholic beverages, many of these assessments have demonstrated that sound can influence the consumer's evaluation of the taste/flavor of wine (North, 2012; Wang and Spence, 2015a), vodka (Wang and Spence, 2015b), beer (Reinoso Carvalho *et al.*, 2016b), and whisky (Velasco *et al.*, 2013). These results suggest that it is possible to produce music that can modulate people's taste and flavor judgments, using the aforementioned literature to underpin the sound stimuli so composed (see Note 1) or chosen (i.e., Hauck and Hecht, 2019; Knöferle and Spence, 2012; Reinoso Carvalho *et al.*, 2015a, 2017). Of particular interest, both Crisinel *et al.* (2012) and Reinoso Carvalho *et al.* (2015a), have demonstrated that the consumer's judgment of the sweetness and bitterness of bittersweet foods (toffee and chocolate, respectively) can be modulated by means of customized 'sweet' and 'bitter' music tracks.

Looking for correspondences across the senses is not the only way in which to trigger associations between what we hear and taste. The fact that a consumer may — or may not — like the music that happens to be playing as part of a multisensory tasting experience, can also exert a significant influence over how the food/drink is evaluated. As such, the multisensory effects that may be observed could be mediated by personal preferences, and/or by the different emotional reactions that the music may induce (e.g., Kantono *et al.*, 2016a; Wang and Spence, 2017, 2018; see Konečni, 2008, for an overview on how music can induce emotion). For instance, Wang and Spence (2017) suggested that certain wine–music pairings are rated as better matches than others, where they argued that wine–music associations are not arbitrary, and can be partially explained by emotional associations.

Different studies have also addressed the question of how different types of sounds and music could influence the subjective value of tasting, potentially adding more pleasure to multisensory eating and drinking experiences (e.g., Fiegel *et al.*, 2014; Kantono *et al.*, 2016b; Reinoso Carvalho *et al.*, 2015c). In particular, Fiegel *et al.* (2014) demonstrated that consumers tend to like the food significantly more while listening to jazz, as compared to hip-hop music. Note that this crossmodal influence of background music was only detected with chocolate (i.e., defined in this study as an emotional food), and not with the type of food pre-defined as non-emotional (bell pepper). In another of these studies, Reinoso Carvalho *et al.* (2015c) assessed whether consumers would agree that music and soundscapes enhanced their tasting experiences. The latter results revealed that not only did the customers report having a significantly better tasting experience when the sounds were presented as part of the food's identity, but they were also willing to pay up to 20% more for such multi-sensory tasting experiences (e.g., involving chocolate–music pairing), when compared to their willingness to pay for the food.

Different studies have assessed the particular influence of sound and music on the behavior of consumers. For instance, Biswas *et al.* (2019) recently reported that people tend to buy healthier food when there is quiet music/noise, as compared to loud background music/noise, or a silent control condition. Moreover, in this study, exposure to loud music/noise led to unhealthier food choices. Meanwhile, Sester *et al.* (2013) reported that the drinks selected by people can be influenced by the semantic, perceptual, or cognitive congruency between such drinks and immersive audiovisual projections of customised bar scenes. Another study assessed whether people's experience of a beer could be enriched by means of audiovisual information (see Reinoso Carvalho *et al.*, 2016c). In the latter study, the presence vs absence of package labeling was manipulated in order to assess the potential effect of the beer's label, and its interaction with music, on the consumer's tasting experience (think of this experiment as an assessment of the potential usage of multisensory packaging solutions in order to, for example, increase brand loyalty). Results show that the beer-tasting experience was rated as more enjoyable with music than when the tasting was conducted in silence. Moreover, those who were familiar with the band who had composed the music track liked the beer more after having tasted it while listening to the music track, than those who knew the band, but only saw the label while tasting.

With the series of experiments reported in the present study, the question of whether different pieces of pre-recorded popular music could be used to modulate the way in which consumers judged different flavor and hedonic attributes of a tasting experience was analyzed. In order to set a suitable and robust methodology, six hypotheses were tested while addressing this question. In contrast to most of the previously-published findings (with a significant portion of them summarized above), here, the experimental music was not chosen based on crossmodal associations (Note 2) (cf. Reinoso Carvalho, 2015a, b, 2016b; Spence *et al.*, 2013; Wang and Spence, 2015a, b, 2016), personal preferences (cf. Kantono *et al.*, 2016a, b, c, 2018), or specific musical genre/style (cf. Fiegel *et al.*, 2014). Whereas most of the existing findings surrounding this topic rely on classifying music based on auditory features (or on somewhat intuitive characterization/composition techniques), here, for the first time, the music was chosen solely on the basis of the emotional state that it would likely induce in consumers. In order to further innovate, and distinguish the way in which the auditory stimuli were selected, the selection of music tracks was validated using a well-standardized and robust method, the Positive and Negative Affection Score — PANAS (cf. Wang and Spence, 2018 [Note 3], where the music tracks were chosen based on their musical consonance vs dissonance; and Wang and Spence, 2017, where the different music tracks were chosen in terms of their tempo, mode, and variability of instrumentation). As far as we are aware, the PANAS has not been used to characterize music with taste/flavor correspondences in mind.

1.1. Theoretical Framework

In these experiments, it was decided to work with different beers as tasting stimuli. Beer is widely consumed at social gatherings. Therefore, it is plausible to assume that music is commonly present in the background of people's everyday beer-tasting experiences, making such experiences most likely under the usual effect of auditory cues (Reinoso Carvalho *et al.*, 2016b). As a general starting point, we assumed that music would somehow affect the evaluation of the tasting experience associated with a drink, as compared to drinking in silence (Bruner, 1990; Cohen *et al.*, 2008).

Hypothesis 1 (H1): Consumers experience a beer differently when there is music playing, as compared to drinking in silence. We continue by focusing on the particular relationship between the beer and the music. In general, consumers tend to rely on their momentary emotional states while evaluating how good or bad — or perhaps how satisfying — an experience/service is (Schwarz and Clore, 1983). We assume, therefore, that some of the sensations and emotional responses elicited by listening to the different music would be transferred into two of the most important dimensions of a tasting experience: hedonic (Hypothesis 2 — H2), and sensory (Hypothesis 3 — H3; see Cheskin, 1972, for an early review of the notion of sensation transference) dimensions. For example, if participants were to prefer one music track over another, such differences might well be expected to affect the corresponding hedonic and sensory dimensions of the tasting experience (Kantono *et al.*, 2016a, b, c). Therefore, if the experimental music were to trigger different emotional states (e.g., positive vs negative), we expected these to mediate the different hedonic aspects of the tasting experience (H2; cf. Noel and Dando, 2015).

Hypothesis 2 (H2): The hedonic experience associated with drinking a beer is affected by the emotions induced by the music that is heard during the tasting experience. Regarding potential effects that music could trigger on the sensory dimension of the tasting experience, music evoking a positive emotion has been shown to enhance the sweetness of different foods and drinks (H3; see Table 1 of Wang and Spence, 2018). For example, Kantono *et al.* (2016a) reported that chocolate ice cream is perceived as tasting sweeter when the music that is playing happens to be liked by the participants. The music tracks used in the aforementioned study were classified along a hedonic dimension, where the participants rated the available music tracks as being liked, neutral, or disliked. Wang and Spence (2018) have also demonstrated that, in the presence of external positively valenced audiovisual stimuli, juice is rated as tasting sweeter, when compared to the same drink when accompanied by negatively valenced external cues, regardless of whether the latter stimuli are visual or auditory.

Hypothesis 3 (H3): The sensory experience of a beer will be affected by the emotions induced by the music that is heard during the tasting experience. It has also been suggested previously that, when looking to design positive experiences for customers, musical congruency in a service setting, or the congruency between the sensory dimensions of a scent and of music, in stores, induces lower arousal — which usually leads to higher pleasure — when compared to corresponding incongruent conditions (Demoulin, 2011; Fiegel *et al.*, 2014; Mattila and Wirtz, 2001). In fact, North *et al.* (1997, 1999) reported that people can be influenced by certain music styles (i.e., varying in ethnicity) while choosing wine. Specifically, playing French music in a supermarket led to higher sales of French wine, while playing German music led to increased sales of German wine instead. More recently, Zellner *et al.* (2017) reported that students ordered more Spanish paella with Spanish music playing in the background of a canteen, as compared to another night where Italian music was played instead. Thus, it was hypothesized that a high level of congruency between the food/drink, and the music, would induce a low level of arousal in the multisensory tasting experience, potentially contributing to the overall perceived pleasantness.

Hypothesis 4 (H4): If the beer and the music do not match in some aspect (sensory or cognitive incongruence), induced arousal will be higher, which would lead to a less pleasant tasting experience, compared to congruent tasting conditions. While choosing and characterizing the different music tracks, even though there was no particular focus on their music genre/style, the implications of differences in genre/style — and how they may affect the study — were still considered (e.g., Areni and Kim, 1993; North *et al.*, 1997, 1999). Previous reports have suggested that people may be willing to pay more for food and drink items when they are accompanied by music (especially in cases where the music is presented as part of a food/drink product's identity; see Reinoso Carvalho *et al.*, 2015c, 2016c). Consequently, in the present study, we also decided to assess the participants' willingness to pay (WTP), based on the idea that the participants might be willing to pay significantly more, or less, for the food/drink, as a function of the music they heard during the tasting experience.

Hypothesis 5 (H5): Participants' WTP for a beer is affected by the music that is being played during the tasting experience (i.e., higher monetary value goes hand in hand with increased hedonic experience). It has often been noted how presenting contrasting combinations of multisensory stimuli typically leads to a more robust observation of crossmodal correspondence effects (i.e., comparing congruent vs incongruent pairings; see Spence, 2011), presumably because it draws attention to the relevant stimulus dimension (see Spence, 2019b). This effect was tested with two different experimental set-ups. In Experiment 1, the same drink was evaluated twice, once in the silent condition and another time while listening to one of the available music tracks. In Experiments 2 and 3, the participants evaluated the tasting experience while listening to two contrasting musical tracks (without a silent baseline condition).

Hypothesis 6 (H6): More robust crossmodal effects should be obtained when the beer is evaluated under the influence of two contrasting music tracks, as compared to evaluating the same beer under the influence of music vs in silence. Following the framework introduced here, one pre-test and three experiments were conducted. In the pre-test, a method of characterizing music based on its ability to evoke different emotions is proposed, using the PANAS scale. In Experiment 1, a multisensory behavioral test was conducted (using the music that was selected in the pre-test), where the participants compared their tasting experience while drinking the same beer twice, once listening to music and the other time in silence. Thereafter, it was decided to test H6 by eliminating the silent control condition of Experiment 1, and using a contrasting pair of music tracks that would evoke either positive or negative emotions (Experiment 2). The latter protocol was further replicated, but now using a very different type of beer, when compared to those used in Experiments 1 and 2, challenging the generalizability of the experimental approach (see Experiment 3).

2. Pre-Test: Musical Selection

The pre-test was designed to support the selection of music that would likely induce different emotional states in participants. The hypothesis of this pre-test was that participants would be able to differentiate music tracks that were chosen to resemble emotions with opposite valence (i.e., positive vs negative).

2.1. Materials and Methods

2.1.1. Participants
A total of 263 participants between the ages of 17 and 75 years took part in the pre-test (52% females, 48% males; mean age = 34 years; SD = 13). The

majority were European residents (69.6% Europeans; 22.4% Americans; 5.7% Asians; 1.1% Oceanic; and 1.1% Africans).

2.1.2. Stimuli

Music was selected that would either evoke positive or negative emotions. It was decided to prioritize strong emotional impact, and, envisioning the applicability potential of these ideas, to work with existing music that was available in the marketplace, and that had been composed by popular/professional musicians. Four music tracks were chosen: Two were chosen as music that would likely be associated with positive emotion (Positive 1 and Positive 2), and the other two as music that would be associated with negative emotion (Negative 1 and Negative 2). Table 1 summarizes the principal characteristics of each of the selected music tracks.

These four pieces of music were cut to approximately one minute each. They were further mastered to roughly equal loudness. The music tracks, as they were played in the pre-test (and in the following experiments as well), can be heard at: https://tinyurl.com/sonictaste-musicandemotions.

In terms of crossmodal correspondences (between sound/music, and basic taste attributes), and following the guidelines provided in Table 1, in Knöferle and Spence (2012), Positive 1 is the only music that may be partially regarded as 'sweet' (since this piece is mainly interpreted by piano melodies, although its frequency range does not fall into the purely 'sweet' category). Positive 2, and Negative 1, on the other hand, do not fall in any particular taste category.

Table 1.
Main characteristics of the experimental music tracks analyzed in this pre-test

Label	Original name of music	Mainly composed by	Technical Summary	Resemblance
Positive 1	Nocturne Op. 9 No. 2	Chopin	E-flat major. This contemplative piece is in rounded binary form (A, A, B, A, B, A) with coda, C, which opens with a legato melody, mostly played by piano. The principal melody repeats three times while, progressively becoming more elaborated, including decorative tones and trills. There is recurrent and considerable rhythmic freedom.	When thinking about Western classical music, this piece is usually regarded as embodying music of the romantic era.

Table 1.

(Continued)

Label	Original name of music	Mainly composed by	Technical Summary	Resemblance
Negative 1	Adagio for Strings	Samuel Barber	Largely in the key of B♭ minor. It is an example of arch-form-building on a melody that first ascends and then descends in stepwise fashion. The lower strings come in two beats after the violins. In general, this composition relies on a tense melodic line with taut harmonies.	This music track is commonly regarded as evoking very negative emotions, and has previously been rated cross-culturally as one of the world's most universally depressing pieces of music (Huron, 2007). It may be familiar to some as the theme music from Oliver Stone's Vietnam-era movie "Platoon" (Note 4).
Positive 2	Porro Sabanero	Lucho Bermudez	The fastest in terms of tempo, and probably the track with the clearest folkloric connotation (Note 5). This music track is called "Porro Sabanero" (by Lucho Bermudez).	The 'porro' is a subgenre of the Colombian Cumbia. Cumbia is interpreted mostly by rich brass bands or orchestras, which are usually accompanied by a strong and rigid rhythmical section (Note 6).
Negative 2	Mors Prae-matura	Jessica Curry	This is a music track that mainly contains a baseline of dissonant-legato orchestral string and brass sections — in the very low-frequency range, when compared to the other three music tracks — along with a high-pitched female voice singing in lyric format.	"Mors Praematura" is part of an ambient soundtrack that is supposed to support a narrative of a horror-indie type of video-game ("Amnesia: A Machine for Pigs").

Moreover, Negative 2 may be regarded as a bitter type of music, due to its aggressive low-frequency baseline. However, the female voice does not allow this music to be framed as purely bitter either. In other words, even though two of the music tracks have clear elements that may position them as a 'sweet' (Positive 1), or 'bitter' (Negative 2), type of music, based on the literature on crossmodal correspondences, none of the four music tracks can be truly framed as purely associated with one particular taste note.

2.2. Experimental Design and Procedure

To validate the different emotions that the musical selection can evoke in participants, the PANAS scale was used (Crawford and Henry, 2004; Watson *et al.*, 1988). This self-report questionnaire consists of 10-item scales designed to measure positive emotional reactions, and 10-item scales to measure negative emotional reactions (i.e., there are 20 items/questions in total). Each answer is rated on a Likert-scale of 1 (not at all) to 5 (very much) — see worksheet 3.1, of Magyar-Moe, 2009, for the template on how the scales were constructed, and how they should be evaluated. For this pre-test, an electronic survey was distributed via the internet, globally, through the usage of different digital sources (e.g., social networks, different mailing lists, etc.). The recruitment process was based on convenience sampling, with some basic filters related to demographics, and technical equipment availability (e.g., the need of headphones or at least a good pair of loudspeakers) (Note 7).

This online survey was subdivided in three parts. First, the participants had to input basic personal data and accept the experimental conditions. They were then informed that they would listen to four musical tracks and, at the end of each one, they should answer a number of questions. In this part, they were also advised that, in order to participate in the survey, they should use headphones set at a comfortable listening level or at least a good pair of loudspeakers. In the second part of the study, the participants had to answer the full PANAS scale, after listening to each of the four music tracks (in total they answered the 20-items questionnaire four times; for each item, the maximum score available was 50, and the minimum 1). They were instructed to answer the questionnaires in terms of how each of the music tracks made them feel. In the third and final part of the study, the participants had to rank the four music tracks in order of preference.

Envisioning the fact that in the following multisensory experiments, the music tracks were to be analyzed in contrasting pairs (e.g., the effects of one positive track would be compared to those of one negative track, and so on), in the pre-test, the music tracks were presented in contrasting pairs as well, with the order of presentation as follows: Positive 1 vs Negative 1 (counterbalanced), and Positive 2 vs Negative 2 (counterbalanced). This means that the presentation of the four music stimuli was not fully randomized. Note that the order of presentation of the questions was fully randomized.

2.3. Results

All of the data analyses reported in the present study were conducted using the statistical package IBM SPSS Statistics (Version 21). Descriptive statistics

Table 2.

Descriptive statistics for positive and negative emotional scores for each music track. Means and corresponding SDs

Emotion score	Music track	Mean (M)	Std. deviation (SD)
Positive score	Positive 1	25.02	7.43
	Negative 1	22.06	8.01
	Positive 2	28.04	8.24
	Negative 2	19.12	6.81
Negative score	Positive 1	12.53	3.39
	Negative 1	15.95	6.35
	Positive 2	12.27	3.31
	Negative 2	23.92	8.66

were calculated following the scoring instructions of the PANAS question-naire (see worksheet 3.1, of Magyar-Moe, 2009), for both the positive and negative emotional scales, and for each one of the music tracks. The higher positive score was for Positive 2 (Mean = 28.04), while the lower positive was for Negative 2 (Mean = 19.12). The highest negative score was for Negative 2 (Mean = 23.92), while the lowest scores were for Positive 1 (Mean = 12.53), and Positive 2 (Mean = 12.27). In general, Positive 1 and Positive 2 clearly evoked more positive than negative emotional reactions, and were the two music tracks with the highest positive ratings. Negative 2 clearly revealed the opposite trend (see Table 2), and Negative 1 and 2 were the two music tracks with the highest negative ratings.

Nevertheless, the results of Negative 1 are not conclusive enough. Although it was positioned more negatively than Positives 1 and 2, it had higher positive than negative scores.

Subsequently, in order to look for significant differences between the four music tracks, a RM-MANOVA was performed for the positive and negative scores, separately (Note 8). The results revealed a significant effect of the music tracks for both the positive ($F(3, 786) = 86.46$, $p < 0.01$, $\eta^2 = 0.25$), and for the negative ($F(3, 786) = 301.62$, $p < 0.01$, $\eta^2 = 0.54$) emotion scores. Subsequently, Bonferroni-corrected post-hoc analyses were conducted (Shaffer, 1995). In terms of positive emotion scores, significant differences were obtained between all of the average means of the music tracks ($p < 0.001$ for all comparisons). In terms of negative emotion scores, there were significant differences between all of the music tracks, except between Positive 1 and Positive 2 (see Table 3 for the comparisons).

Finally, the descriptive analysis regarding the way in which the participants ranked the music tracks in terms of their preference revealed that Positive 1 was principally ranked as the favorite track (45% of the time), Positive 2 was

Table 3.

Differences between the means of the music tracks (asterisk '*' indicates a significant difference at $p < 0.001$). In the positive emotion scores, we see that the differences between Positive 1 and Negative 2, between Positive 2 and Negative 2, and between Negative 1 and Positive 2 are more pronounced, as compared to the other mean differences. When it comes to the negative emotion scores, the differences between Positive 1 and Negative 2, and between the Positive 2 and Negative 2, stand out from the rest

Music track	Positive emotion score			Negative emotion score		
	Negative 1	Positive 2	Negative 2	Negative 1	Positive 2	Negative 2
Positive 1	2.88*	2.95*	5.92*	3.46*	.23	11.41*
Negative 1		5.84*	3.03*		3.69*	7.95*
Positive 2			8.87*			11.64*

mostly ranked as the second favorite (32% of the time), Negative 1 was mostly ranked as the third favorite (42% of the time), and Negative 2 was mostly ranked as the least favorite (59% of the time).

In summary, three of the music tracks exhibited clear expected effects (Positive 1, Positive 2, Negative 2). The two positive music tracks were clearly ranked higher relative to the positive emotional PANAS score (see Tables 2 and 3). Regarding the corresponding negative emotional score, Negative 2 ranked higher by far. Interestingly, the usage of PANAS in the pre-test was well-founded, since it allowed us not only to validate the choice of the music tracks, but also to find one music track that did not behave as expected. Even though Negative 1 was initially framed as an emotionally negative type of music, after the corresponding PANAS analysis, it did not present fully conclusive effects. Hence, it was decided not to use Negative 1 in the following experiments.

Having successfully selected the music, in the following experiments we went on to assess whether the emotional reactions induced by these selected music tracks would influence specific hedonic, sensory, preference, and WTP aspects of a consumer's tasting experience.

3. Experiment 1: Tasting Beer With Music vs in Silence

In Experiment 1, the participants sampled a beer while listening to the music selected in the pre-test, and evaluated their experience by means of a questionnaire. Here, we analyzed whether the emotional reactions induced by these musical selections would influence specific hedonic and sensory aspects of the beer tasting experience, as well as preference and WTP responses.

3.1. Materials and Methods

3.1.1. Participants
A total of 221 new participants took part in Experiment 1 (45% females, 55% males; 85% European residents). The mean age was 32 years (range 16–69 years, SD = 13) (Note 9).

3.1.2. Stimuli
3.1.2.1. Flavor Stimuli. In Experiment 1, the beer used was 'Zinnebir' (produced by Brasserie de la Senne, a small Belgian brewery). It is a Belgian bitter-dry pale lager, with 6% alcohol (malty, with fine-salient bitterness (Note 10). The beers were served in normed 10 cl samples (in order to avoid satiation), and in opaque black plastic cups (in order to prevent the participants from basing their responses on the beer's color).

3.1.2.2. Auditory Stimuli. Two of the three available music tracks were used: Positive 1 and Negative 2 (meaning one positive and one negative, respectively). All listening systems were calibrated to have approximately the same sound pressure level (L_{eq30s} of approximately $70+/-3$ dBA).

3.2. Experimental Design and Procedure

This experiment was conducted in an isolated area, where it was possible to have a fairly controlled environment during testing hours (Note 11). Each participant was seated in front of a computer screen with a pair of headphones, a computer mouse, and a keyboard with which to complete the survey. The music was presented over Sony MDRZX-310 headphones. The survey consisted of an electronic form containing three main steps. First, the participants were instructed to read and accept the conditions of the informed consent before entering their demographic details. In the second step, the participants tasted two identical samples of the same beer in two different trials, without being told that they were, in fact, tasting the same beer. The participants tasted the beer once while listening to either positive or negative music, and once in silence. After tasting each beer, they had to answer questions related to their hedonic and sensory experience. These answers were reported by means of 7-point rating scales. During the experiment itself, the beers were labeled as TK (when tasted while listening to music), and WD (when tasted in silence). The order in which the questions were presented was randomized. The order of the sound condition (music vs silent condition) was counterbalanced across participants as well. The third and final part of this questionnaire contained complementary multiple-choice and YES/NO questions related to the participants' beer preferences and price judgment. The presentation of these questions (and hence the corresponding choices of answers) was randomized (Note 12).

Table 4.

Means and SD of ratings related to the participants' beer evaluation, for hedonic and sensory ratings in Experiment 1. The first two rows are a comparison of the ratings of participants on the silent condition (first row), and music condition (second row; negative + positive music effects). The third and fourth rows are a comparison of the results of the music condition, disentagling negative vs positive effects. The results that are highlighted in bold show a significant difference between the two corresponding conditions

Group	Beer liking	Sweetness	Bitterness	Sourness	Alcohol strength	Body
Silent condition	4.30 (1.57)	2.81 (1.37)	4.64 (1.34)	3.55 (1.44)	4.03 (1.25)	**3.81 (1.28)**
Music condition	4.34 (1.66)	2.71 (1.30)	4.80 (1.37)	3.65 (1.57)	4.24 (1.25)	**4.07 (1.22)**
Negative music	4.18 (1.48)	2.65 (1.25)	4.73 (1.29)	3.59 (1.54)	4.11 (1.19)	3.86 (1.31)
Positive music	4.49 (1.74)	2.89 (1.34)	4.71 (1.46)	3.61 (1.61)	4.17 (1.32)	4.03 (1.11)

3.3. Results

Hedonic (e.g., liking of beer), sensory (e.g., flavor attributes of beer, such as sweetness, and alcohol strength), and other preference/consumer ratings (e.g., WTP) were analyzed as dependent variables. In general, the mean scores tended to be higher in the music condition than in the silent condition (see Table 4). In particular, the beer received higher ratings related to the flavor attributes of the beer (except for sweetness). For liking ratings, the average music score (negative + positive music effects) came close to the average of the silent condition score. For liking, the positive music received higher ratings than the negative music, with the silent condition in-between them.

A RM-ANOVA was further conducted for each of the two experimental conditions (silent vs positive, and silent vs negative), and for each scale, separately (Note 13). Bonferroni correction was applied for multiple comparisons (confidence interval, $\alpha = 0.05/3 = 0.017$; see Cramer *et al.*, 2016).

When asked how much they liked each beer, there were no differences observed. In terms of the perceived alcohol strength of the beer, a within-participants trend was observed (silent $M = 4.03$–music $M = 4.24$) $[F(1, 219) = 5.40, p = 0.021, \eta^2 = 0.024]$, suggesting that people may have rated the beer as tasting stronger (i.e., more alcoholic) while listening to music, regardless of the type of music (note that a similar result has been obtained previously in the case of fine wine tasted either with or without music; see Spence *et al.*, 2013, Experiment 3). For ratings of the beer's body, there was a main effect for within-participants (silent $M = 3.81$–music $M = 4.07$) $[F(1, 219) = 8.22, p = 0.005, \eta^2 = 0.036]$. The latter result means that in the music condition, and regardless of the type of music that the participants were listening to, the beer was rated as having more body when compared with the silent condition. No differences were found between the other sensory ratings of the beer (sweetness, bitterness, and sourness).

An additional independent t-test was conducted in order to assess how much the participants liked each of the music tracks. Significant differences were found between the music groups [$t(219) = 11.11$, $p < 0.001$], where the participants liked the positive music more ($M = 5.50$) than the negative music ($M = 3.31$). Moreover, each participant had to choose whether they preferred to consume the beer while listening to music, or in silence. More participants preferred to consume the beer while listening to music, no matter whether the music was positive or negative (67% and 55%, respectively) (Note 14). The participants also rated how much they thought each of the music tracks matched the flavor of the beer. However, no significant effects were observed.

In summary, the results of Experiment 1 demonstrate that listening to the music exerted a few effects on people's rating of the beer. Here, the beer was rated as having more body (and was rated as potentially stronger in terms of alcohol content) when tasted with music, as compared with the ratings obtained in the silent condition (providing some support to H1).

4. Experiment 2: Tasting Beer With Positive vs Negative Music

In Experiment 2, the silent control condition was omitted and, instead, just the music having positive and negative emotion was presented to each participant (see H6). The participants were also subdivided into two groups. Each group now listened to two different pairs of contrasting music tracks. Moreover, a different beer was used, in order to investigate whether the different flavor attributes of the different beers would play a significant role in such multisensory effects.

4.1. Materials and Methods

4.1.1. Participants
A total of 154 new participants (54% females, 46% males; 92% European residents) took part in Experiment 2. The mean age was 33 years (range 17–69 years, SD = 13).

4.1.2. Stimuli
4.1.2.1. Flavor Stimuli. The most popular and most consumed Belgian pale-lager beer (5.2% alcohol) was used in Experiment 2 (Jupiler, produced by Anheuser-Busch InBev) (Note 15).

4.1.2.2. Auditory Stimuli. Three music tracks from the pre-test were presented in contrasting pairs to the participants: Positive 1 vs Negative 2 (from now onwards, renamed simply as Negative), and Positive 2 vs Negative.

Table 5.

Means and SD (SD values in parenthesis) of ratings related to the participants' beer evaluation for hedonic and sensory ratings in Experiment 2. The first two rows provide a comparison of the ratings of participants while listening to the Negative vs Positive (Positive 1 + Positive 2) music. The third and fourth rows are a comparison of the ratings of participants while listening to the Positive 1 vs Positive 2 music. Results highlighted in bold show a significant difference between the two corresponding conditions

Music condition	Liking	Sweetness	Bitterness	Sourness	Alcohol strength	Body
Negative	**4.17 (1.55)**	**2.60 (1.25)**	**4.34 (1.33)**	3.55 (1.53)	**4.10 (1.33)**	**4.06 (1.38)**
Positive	**4.81 (1.27)**	**3.28 (1.3)**	**3.79 (1.37)**	3.36 (1.43)	**3.62 (1.35)**	**3.66 (1.29)**
Positive 1	4.56 (1.45)	3.03 (1.29)	**4.28 (1.25)**	3.45 (1.39)	4.05 (1.37)	3.88 (1.38)
Positive 2	4.41 (1.37)	2.86 (1.31)	**3.84 (1.42)**	3.46 (1.57)	3.66 (1.29)	3.85 (1.25)

4.2. Experimental Design and Procedure

Briefly, the experimental design and procedure were the same as in Experiment 1, with a few variations, as follows:

1) The silent control condition was replaced by the presentation of two contrasting music tracks.

2) The beer was switched from Zinnebir to Jupiler.

3) The participant's WTP was assessed by means of a 7-point scale (presented in Euros).

4) Beer labels were WD (negative music), and TK (positive music).

4.3. Results

All of the average mean scores related to liking ratings were higher for the beer when combined with positive music than when combined with negative music. The same was also true in terms of sweetness ratings. By contrast, bitterness, alcohol strength, and body ratings revealed the opposite trend, with the negative music resulting in higher scores than the positive music (see Table 5).

A RM-ANOVA was conducted for each of the two experimental conditions (Positive 1 + Negative; Positive 2 + Negative), and for each scale, separately (positive music tracks together vs negative music track).

4.3.1. Hedonic Beer Evaluation

When asked how much they liked each beer, a main within-participants effect was observed ($[F(1, 152) = 19.343, p < 0.001, \eta^2 = 0.113]$), where participants reported liking the beer more while listening to the positive music ($M = 4.81$) whilst while listening to the negative music ($M = 4.17$).

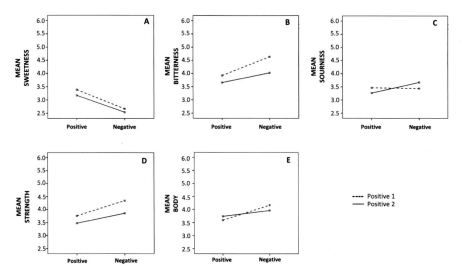

Figure 1. Interaction effect in repeated-measures analysis, for the sensory evaluation of the beer in Experiment 2. The *X*-axis represents the within-participants analysis (positive vs negative music). The *Y*-axes represent the estimated marginal mean ratings of each flavor attribute — e.g., 'A' for sweetness, and 'D' for alcohol strength (with the full-scale going from 1 to 7). Positive 1 results are represented with dashed lines, whereas Positive 2 results are represented with solid lines.

4.3.2. Sensory Evaluation of the Beer (Flavor Attribute Ratings)

A main within-participants effect on sweetness ratings was observed [$F(1, 152) = 27.22$, $p < 0.001$, $\eta^2 = 0.152$], with the participants rating the beer as tasting sweeter while listening to the positive music ($M = 3.28$) than while listening to the negative music ($M = 2.60$; see Fig. 1A).

A main within-participants effect was observed for bitterness (positive music $M = 3.79$–negative music $M = 4.34$; [$F(1, 152) = 15.57$, $p < 0.001$, $\eta^2 = 0.093$]). Specifically, the participants rated the beer as tasting more bitter whilst listening to the negative as compared to the positive music. There was a main effect of group (Positive 1, $M = 4.28$–Positive 2, $M = 3.84$; [$F(1, 152) = 7.02$, $p = 0.009$, $\eta^2 = 0.044$]), where the participants rated the beer as tasting more bitter while listening to Positive 1 than while listening to Positive 2 (see Fig. 1B).

In terms of the rated alcohol strength, there was a main within-participants effect (positive music $M = 3.62$–negative music $M = 4.10$; [$F(1, 152) = 14.650$, $p < 0.001$, $\eta^2 = 0.088$]), with participants rating the beer as tasting stronger while listening to the negative music, as compared to their ratings while listening to the positive music (see Fig. 1D).

A main within-participants effect for the ratings of the beer's body was also observed [$F(1, 152) = 11.496$, $p = 0.001$, $\eta^2 = 0.070$], with the beer being rated as having more body while the participants listened to the negative music ($M = 4.06$), when compared to the ratings related to the positive music ($M = 3.66$ — see Fig. 1E). Summarizing, the obtained results related to the sensory evaluation of the beer supports H3.

4.3.3. Preference and WTP

A main within-participants effect was observed for music preference ($[F(1, 152) = 170.19$, $p < 0.001$, $\eta^2 = 0.528]$), with higher ratings for the positive music ($M = 5.42$) as compared to the negative music ($M = 3.24$). There was also a trend for an effect of group (Positive 1, $M = 4.51$–Positive 2, $M = 4.15$; $[F(1, 152) = 4.44$, $p = 0.037$, $\eta^2 = 0.028]$), suggesting that the participants preferred Positive 1 over Positive 2. An interaction effect between repeated measures (positive–negative music) and group (Positive 1–Positive 2) $[F(1, 152) = 6.94$, $p = 0.009$, $\eta^2 = 0.044]$, was also observed, which confirms that there were clear differences on the way the participants appreciated each music track.

The participants reported liking the positive music more than the negative music. The participants were also asked to estimate their WTP for each of the two beers, in Euros. An additional dependent t-test was conducted to compare the WTP ratings for the TK beer (Jupiler combined with positive music), and for the WD (Jupiler combined with negative music). There were significant differences (Note 16) between both ratings $[t(153) = 2.055$, $p \sim 0.004$; TK Beer $= 4.29$ Euros; WD Beer $= 4.00$ Euros], showing that the participants were willing to pay 7.25% more for the Jupiler beer, when tasted while listening the positive music, as compared to while listening to the negative music (supporting H5). The participants were further asked to choose which beer they preferred (TK, WD, or no preference). A higher percentage of participants preferred the TK beer (Jupiler combined with positive music), in both groups (Positive 1 $= 58\%$, Positive 2 $= 49\%$) (Note 17).

In summary, the results of Experiment 2 highlight more pronounced differences between the effects of the different music (positive vs negative) on the multisensory beer tasting experience, when compared to the effects that were reported in Experiment 1 (supporting H6). These results also suggest that the beer ratings may be under the influence of the different emotional reactions triggered by the music (the kind of influence that has been referred to as 'sensation transference' by Louis Cheskin, e.g., Cheskin, 1972).

5. Experiment 3

In the final experiment, we decided to try and replicate the promising results obtained in Experiment 2 while changing the beer that the participants evaluated (e.g., Watson and Gunther, 2017; see Open Science Collaboration, 2015, on the importance of replicability in psychological — and related/applied — sciences).

5.1. Participants

A total of 157 new participants took part in this experiment (48% females, 52% males; 94% European residents). The mean age was 35 years (range 16–70 years, SD = 14).

5.2. Experimental Design and Procedure

The experimental design and procedure of Experiment 3 were the same as for Experiment 2, except for the type of beer. Chimay Blue was chosen as the new beer because it is very different from the types of beer that were used in the previous experiments. This is a Belgian strong dark ale type of beer, with 9% alcohol. This dense beer has a strong caramel flavor, and a smooth palate sensation (Note 18).

5.3. Results

Reassuringly, in general, the results exhibit similar trends to those reported in Experiment 2 (except for the ratings of the beer's body). The ratings of liking and sweetness were higher for the positive music, when compared to the ratings given while listening to the negative music. On the other hand, bitterness, alcohol strength, and ratings of body were higher when it came to the ratings related to the negative music (see Table 6).

5.3.1. Hedonic Evaluation of the Beer
A main within-participants effect was observed for beer preference [$F(1, 155) = 12.02$, $p = 0.001$, $\eta^2 = 0.072$], with participants liking the beer more

Table 6.
Means and SD of ratings related to the participants' beer evaluation in Experiment 3. The first two rows are a comparison of the ratings of participants while listening to the Negative vs Positive (Positive 1 + Positive 2) music conditions. The third and fourth rows provide a comparison of the ratings of participants while listening to the Positive 1 vs Positive 2 music conditions. Results highlighted in bold show a significant difference between the two corresponding conditions

Music condition	Liking	Sweetness	Bitterness	Sourness	Alcohol strength	Body
Negative	**4.22 (1.51)**	**3.19 (1.32)**	**4.25 (1.34)**	3.67 (1.37)	**4.50 (1.29)**	4.38 (1.24)
Positive	**4.79 (1.63)**	**3.67 (1.39)**	**3.88 (1.37)**	3.59 (1.49)	**4.20 (1.25)**	4.16 (1.26)
Positive 1	4.59 (1.53)	3.39 (1.31)	4.03 (1.33)	3.52 (1.44)	4.43 (1.26)	4.36 (1.20)
Positive 2	4.40 (1.60)	3.46 (1.43)	4.09 (2.12)	3.75 (1.54)	4.27 (1.29)	4.18 (1.32)

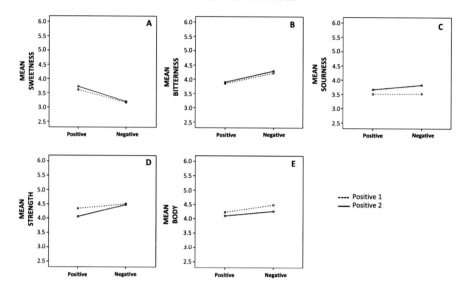

Figure 2. Interaction effect in repeated measures analysis, for the sensory evaluation of the beer in Experiment 3. The X-axis represents the within-participants analysis (positive vs negative music). The Y-axes represent the estimated marginal mean ratings of each flavor attribute — e.g., 'A' for sweetness, and 'D' for alcohol strength (with the full-scale going from 1 to 7). Positive 1 results are represented with dashed lines, whereas Positive 2 results are represented with solid lines.

accompanied by the positive music ($M = 4.79$) than when accompanied by the negative music ($M = 4.22$).

5.3.2. Sensory Evaluation of the Beer

A main within-participants effect on sweetness ratings was observed [$F(1, 155) = 13.01$, $p < 0.001$, $\eta^2 = 0.077$], with participants rating the beer as tasting sweeter while listening to the positive music ($M = 3.67$) as compared to the negative music ($M = 3.19$) — see Fig. 2A.

A main within-participants effect on bitterness ratings was observed [$F(1, 155) = 7.32$, $p = 0.008$, $\eta^2 = 0.045$]), with the beer being rated as tasting more bitter while listening to the negative music ($M = 4.25$) as compared to the positive music ($M = 3.88$) — see Fig. 2B. For beer alcohol strength, a main effect for within-participants was observed [$F(1, 155) = 6.32$, $p = 0.013$, $\eta^2 = 0.039$]), with the participants rating this beer as tasting stronger while listening to the negative music ($M = 4.50$) as compared to the positive music ($M = 4.20$) — see Fig. 2D.

5.3.3. Preference and WTP

When asked how much they liked the music, a main within-participants effect was observed [$F(1, 155) = 139.88$, $p < 0.001$, $\eta^2 = 0.474$], with participants liking the positive music more ($M = 5.32$), than the negative music ($M =$

3.22). There was also a main effect of group [$F(1, 155) = 7.33$, $p = 0.008$, $\eta^2 = 0.045$], with participants reporting that they liked Positive 1 ($M = 4.48$) more than Positive 2 ($M = 3.99$).

The corresponding WTP ratings show significant differences between the beers with a dependent t-test [$t(156) = 2.17$, $p = 0.031$; TK Beer $= 4.29$ Euros; WD Beer $= 4.00$ Euros]. The same difference in price was observed as in Experiment 2, where the participants were willing to pay 7.25% more for a beer while being tasted under the influence of the positive as compared to the negative music. A higher percentage of participants also preferred the TK beer (consumed with positive music) in both conditions (Positive 1 $= 49\%$, Positive 2 $= 46\%$) (Note 19).

In general, Experiment 3 works as a validation of the methodology proposed in Experiment 2. The results are very similar, except for the ratings of the beer's body. In Experiment 2, the Jupiler beer was rated as having significantly more body when tasted under the influence of the negative music, when compared to the ratings related to the positive music. In Experiment 3, the trends of body ratings were similar, but they failed to achieve statistical significance. (See supplementary material for complete dataset.)

6. General Discussion

In these experiments, the effects of listening to positive vs negative emotion type of music on the consumer's tasting experience were studied. We investigated whether different pieces of popular music could be used to modulate the way in which consumers judge the experience of drinks (in this case, a beer). In terms of methodology, the principal novelty that this study brings to this topic is that the experimental music was not chosen based on the particular auditory features *per se* (Note 20), but rather — and solely — based on the emotional state that such music could induce on the consumers. In general, the results suggest that the emotional reactions triggered by the music came to influence specific aspects of the consumer's tasting experience. The results of Experiment 1 suggest that music can affect the way consumers rate certain aspects of a tasting experience (e.g., a beer's perceived body), when compared to drinking it in silence (regardless the type of music; H1). The results of Experiments 2 and 3 provided support to most of the research assessments tested here. In particular, it was shown that the hedonic (H2) and sensory (H3) experiences of a drink can be affected by the emotions induced by the music that is being played during the tasting experience (including a consumer's WTP for a drink — H5). Hence, Experiments 2 and 3 highlighted more pronounced differences between the effects of the different music (positive vs negative) on the multisensory tasting experience, when compared to the effects reported in Experiment 1 (giving support to H6, which states that more robust crossmodal effects tend to be reported when a tastant is evaluated under the influence of two contrasting music tracks, in comparison to evaluating the influence of one of the available music tracks vs in silence). Another complementary — and foremost — part of our research concerns the way in which the music should be chosen for this type of exercises. Here, we proposed the PANAS as a reliable way to validate such musical choices (at least, when it comes to understanding the emotional reactions that such music may elicit in people — e.g., positive or negative emotions). Importantly, the usage of PANAS in the pre-test was well founded, since it allowed us not only to validate the intuitive choice of the music tracks, but also to detect one music track that did not behave as expected.

As the main contribution of this article, we demonstrate that it is possible to implement a robust-replicable behavioral — and quantitative-based — methodology, using popular music (classified as evoking positive or negative emotions) to modulate the way in which consumers judge different aspects of a multisensory tasting experience (including sensory attributes of the beer, preference, liking ratings, and WTP). Such effects are clearly detected when two contrasting music tracks (in terms of positive vs negative emotions) are presented to the consumer, while, in this case, tasting the same beer. This novel methodology is unique because it allowed us, for the first time, to clearly disentangle the effects that the emotional mediation elicited by music is bringing to a tasting experience. All of this, regardless of how the music was composed, or characterized, in terms of its sonic/musical attributes.

In the pre-test, three out of the four music tracks tested had the expected effects (Positive 1, Positive 2, and Negative 2; see Table 3). The results related to Negative 1 were, however, not so clear-cut. The latter may have to do with the way that this music track had been cut. Perhaps the fragment of Negative 1 that was chosen for this exercise was not representative enough in terms of negativeness.

In Experiment 2, most of the hedonic, sensory, and consumer evaluations of the beer were clearly influenced by the different emotional reactions that the positive and negative music had on the participants' beer experience (see Fig. 1, and compare with the results of Experiment 1). The emotional reactions triggered by the music were somehow transferred into specific hedonic and sensory aspects of the beer-tasting experience. For instance, the beer was liked more, and it was judged as tasting sweeter while listening to the positive music (see H2 and H3). On the one hand, there is positive emotion induced by this music. On the other hand, the positive music was preferred over the negative music. Both aforementioned effects support a positive sensation transference (Cheskin, 1972), which boosted pleasure, thus most likely activating the reward system (by providing a more enjoyable tasting experience that, as a consequence, was rated as tasting 'sweeter') (Note 21).

Moreover, in Experiment 2, the participants also reported being willing to pay 7–8% more for the same beer, while tasting it under the influence of the positive music, when compared to how much they were willing to pay under the influence of the negative music (see H5). This could also be related to the positive transference of sensations explained above.

In contrast, the beer was judged as tasting more bitter, as being more alcoholic, and as having more body, when experienced under the influence of the negative music (see H3). In this case, the negative music brought negative emotions to the tasting experience. It has been suggested that people in negative emotional states tend to search for more information in order to explain/resolve their mental condition, when compared to when in a more pleasant situation (Schwarz and Clore, 1983). Therefore, it could be argued that such a negative emotional state mostly resulted in less attention being devoted towards the music, and more attention being directed towards the most salient aspects of a beer's taste/flavor, such as the corresponding bitterness, and alcohol percentage (as in a kind of attentional shift/redirection effect; see Johnson and Proctor, 2004; Spence, 2014). Interestingly, the fact that the most salient flavor characteristics of the beer were enhanced by the mediation of the least preferred — and negatively-valenced type of — music, suggests that, depending on the desired outcomes, negativeness and incongruent multisensory associations may be as useful as the opposite — and perhaps more common — strategy.

The results of Experiment 3 replicated those of Experiment 2 while changing the beer evaluated by the participants. These new results were very similar to those reported in Experiment 2 (see Table 6; cf. Figs 1 vs 2). The only rating that did not achieve statistical significance in our final experiment was the difference in terms of the judgment of the beer's body (this difference achieved statistical significance in Experiment 2). The reason why this rating did not deliver the expected results might have to do with the fact that rating a drink's body tends to be a more complex type of evaluation for most people (e.g., when compared to judging its sweetness or bitterness), especially as far as naïve drinkers are concerned. In light of this finding, it may be interesting in future similar assessments to compare ratings between participants in terms of their level of experience/expertise with regard to the experimental tasting stimuli (e.g., in this case, their experience/expertise as beer drinkers).

The music tracks used in these experiments were musically and ethnically different. Obviously, they were also different in terms of their ability to evoke different emotions, as validated by the results of the pre-test. All of these differences triggered a clear contrast in terms of musical preference. However, such dissimilarities did not result in a significant difference in terms of a better/worse match between the music and the beer(s) (Note 22). As such, with these results it is not possible to draw a conclusion regarding the importance of the potential arousal induced by the music in the beer experience (that is, there is no support for H4). Nevertheless, it can be argued that the effects tested here are mostly triggered by differences in valence, whereas potential arousal effects do not seem to be salient enough. The latter results would also appear to suggest that transfer effects are not generic, but they can be specific to valence, or arousal (cf. Fritz *et al.*, 2017; Marin *et al.*, 2017).

Summarizing briefly, a suitable methodology can emerge from the combination of the protocol presented in the pre-test with the protocol implemented in Experiments 2 and 3 (see H6). Such a novel methodology, as outlined here, can help better standardize the quantification of the effects that emotions induced by music can have on a consumer's drinking experience. We anticipate that the effects reported here with very different beers would be similar for different drinks, and perhaps even for different foods. On top of that, from this methodology, one could expect similar effects of other kinds of emotion-inducing stimuli. All of this could be further tested with different sensory/stimulatory combinations, such as touch, vision, while combined with different artistic expressions, and not only music.

6.1. Implications for Practitioners

Most brands tend to focus on the construction of a strong visual identity, without considering how, for example, sound and music can be used as a relevant attribute, with positive impact in a brand's awareness, and preference (Arora and Kumar, 2018). Hence, and given the results obtained here, we argue that music can be used to add value to the way in which a consumer responds to a brand (at least in the case of beers, as tested in this study). In this sense, the emotions induced by music can also potentially support a brand's positioning. For example, music that induces positive emotions could be an easy way to positively engage the consumer. On the other hand, and as mentioned above, people in negative emotional states tend to search for more information in order to explain their mental condition (Schwarz and Clore, 1983). Therefore, music that induces negative emotion should not be considered as necessarily having a negative impact on the experience of consumers. On the contrary, a positive experience can also arise for consumers, where negative music could be delivered as a type of music that induces more complex emotional reactions, which can result in more attention being paid to, for instance, the particular taste/flavor notes in a food/drink. Hence, those companies looking to explore different competitive positioning strategies based on positioning mapping techniques (e.g., sensory vs hedonic dimensions) could widen their scope of choices by focusing on more complex sonic identities as part of the brand experience (e.g., luxury branding/retailing). Overall, these results (which are mostly triggering effects in terms of consumption, and potentially choice) can also be taken to argue towards the usability of music as a tool to improve brand equity, which may ultimately help drive in increasing market value (Aaker and Biel, 2013).

Actually, global food and drink companies already seem to rely on these type of multisensory experiential design techniques as part of their branding strategies when, for example, developing events, or as support for advertising campaigns (e.g., Campari [Note 23], and Godiva [Note 24]). Recently, for instance, the city of Brussels (Belgium) funded a project entitled 'The Sound of Chocolate' (Note 25), where chocolate boxes were sold alongside music tracks designed to enhance certain aspects of the chocolate's taste and flavor. Besides proposing a new way of experiencing Belgian chocolate, through music, this project had the parallel aim of 'branding' Brussels as an innovative city. Similar multisensory experiences are also being offered by talented music composers, such as Maxime Goulet (Note 26). Ideas such as these may be useful when thinking of experiential retail. For instance, supermarkets may soon start to significantly reduce their physical spaces (following the expansion of e-commerce), and focus on more meaningful in-situ experiences. Consumers that decide to come personally to supermarkets could further engage through similar types of multisensory tasting experiences, while looking to learn about new products and their possibilities.

Nevertheless, the aforementioned existing examples (and a great portion of the existing scientific research surrounding this topic) rely mostly on classifying music based on auditory features (e.g., sonic seasoning), or on somewhat intuitive characterization/composition techniques. With these new results, we stress the importance of considering the emotional mediation of music in multisensory tasting experiences, along with the implications of the personal musical preferences of the consumers. These new results provide further support with respects to the need of more controlled and standardized multisensory food/drink–music pairing methods. As such, personal music playlists could be classified as a function of tastes/flavors, and potentially delivered to the end user by means of online streaming companies, such as Spotify, Apple Music, or Amazon music (Note 27). The further impact of streaming technology on retailing and consumer services may allow consumers to be aware, from their own musical preferences, which type of music would be better to listen to, while eating or drinking specific products (Reinoso Carvalho *et al.*, 2016b; Velasco *et al.*, 2016). Eventually, such multisensory experiences may also be explored along with novel media trends, such as 'mulsemedia' solutions, where multisensorial media delivery is being shown to outperform existing multimedia delivery solutions, in both, perceived quality, and enjoyment (Yuan *et al.*, 2015).

6.2. Limitations

Music often provides changes in mood during an entire track. Therefore, it may be challenging to comprehend how the emotions prompted by a specific musical fragment will develop throughout a tasting experience that involves entire music tracks. In this sense, one should keep in mind that consumers also tend to rely on initial external cues while understanding the characteristics of flavor (and while keeping it homogeneous over time; Woods *et al.*, 2010). Hence, it may be assumed that the judgment of the drink will most likely be set along with the first — or, perhaps, along with the highlighted — musical moment of the multisensory tasting experience.

When naïve consumers are asked to judge not-so-obvious flavor notes, such as the body of a beer, there is no certainty that such consumers are fully aware of what he/she is actually aiming for, especially when there is no reference. Perhaps one way to tackle the latter in future research would be to provide an initial reference point (e.g., start by rating the body of a neutral drink, such as water, prior the evaluation of the body of the experimental tasting stimuli).

The PANAS scale, as it is meant to be used, does not allow a more precise subdivision of the emotional reactions that the music is bringing into the tasting experiences conducted here (e.g., a clearer differentiation between, say, anger vs sadness, both different reactions that, nonetheless, fall within the same negative emotional dimension). Therefore, a way to complement this study comparing contrasting positive and negative emotional dimensions would be to compare positive (or negative) tracks that are somehow complementary/incompatible within their respective emotional dimensions (e.g., compare the effects of two sad music tracks, vs the effects of one sad vs one angry music track, and so on). With such type of comparisons, it may be possible to better disentangle the way in which emotions mediate such multisensory effects.

Finally, some may argue that these types of experimental design may be susceptible to demand, or social desirability, effects (see Rubin, 2016, for an understanding of demand effects, and/or Edwards, 1957, for an understanding on social desirability effects). Therefore, future assessments could install different ways to mitigate such potential confounds, for instance, by trying to 'camouflage' the hypothesis (e.g., 'mask' the connection between the music and the food/drink at stake).

Author's Contributions

FR, JW, and CS designed the study. FR led the data collection, and the selection/design of the different experimental stimuli. SD led the data analysis. All of the authors participated in the preparation of the manuscript and revised the final version of the manuscript.

Funding

FR and JW were supported by KU Leuven IOF internal funding scheme (C32/17/005). FR was also individually supported by Universidad de los Andes' FAPA internal funding scheme (FAPA N.32). JW was also supported by the Flemish Methusalem program (METH/14/02).

Ethics Statement

This study was carried out in accordance with the recommendations of the Social and Societal Ethics Committee at KU Leuven (SMEC, protocol registered as G-2015 07 281). All of the participants gave their written informed consent in accordance with the Declaration of Helsinki.

Acknowledgements

We would like to thank La Brasserie de la Senne for donating the beers used in Experiment 1. We would also like to acknowledge the support of Prof. Raymond van Ee during the first stage of the experimental design, and Dr. Pieter Moors for initial insights regarding the first batch of collected data. We would further like to thank François Nelissen, Elien Haentjens, Sarah Ahannach, Sam Van Broeck, Dr. Rebecca Chamberlain, Carlos Garcia, Midas Vanooteghem, Sarah Delcourt, Shyam Sekaran, Eleftheria Pistolas, Robin Panassie, Beyza Ozen, Aysun Duyar, Elisabeth Vanderhulst, Anurag Suri, Tereza Buckova, and Charlotte Buhre for assistance during data collection, and a very special thanks to the mim (Musical Instruments Museum Brussels), for hosting these experiments, and collaborating during the experimental days.

Supplementary Material

Supplementary material is available online at:
https://brill.figshare.com/s/c57e90299b6f5c268288

Notes

1. In the study reported by Reinoso Carvalho and colleagues (2015a), three music tracks were produced, one designed to be congruent with sweetness, another with bitterness, and the third lying somewhere in-between.

2. Also known as 'sonic seasoning'; e.g., music classified by its sonic and musical characteristics, where such characteristics can be associated with specific taste or flavor attributes (see Spence, 2017c).

3. Note that in this study musical dissonance was presented as an inducer of negative emotion, and vice versa. However, the effects reported were regardless of whether the stimuli were visual or auditory.

4. http://www.imdb.com/title/tt0091763/, retrieved April, 2018.

5. When thinking about ethnic congruency, it could be argued that this music track, as compared to the others used in this pre-test, could be rated as very incongruent when experienced with beer.

6. This genre is also regarded as being influenced by some of the Latin-American bands of the 1960's, and it is primarily meant for dancing.

7. The survey was provided in English, French, Portuguese, and Spanish. The only country where this survey could not be carried out was Germany, since they have strong playback filters that do not allow consumers to reproduce fragments of music (e.g., via Youtube, with Youtube being the streaming platform here used) that may have some relevance in terms of intellectual property disputes.

8. This statistical method was adopted given the fact that each participant evaluated the entire stimulus set, and more than one dependent variable was measured (Huberty and Olejnik, 2006).

9. The minimal legal age for drinking beer in Belgium, the country where these experiments took place, is 16 years of age. Moreover, in order to determine the sample size, a power analysis was performed based on Friedman's simplified determinations of statistical power (see Friedman, 1982, Table 1). Considering 95% confidence ($\alpha = 0.05$), effect size of 0.25, and a power effect of at least 0.8, the suggested sample size would be 120 participants.

10. Extracted from http://brasseriedelasenne.be/?portfolio=zinnebir, April, 2018.

11. There were four rectangular tables where eight participants could enter at the same time (two per table, with a total of eight computers running at once). The experience was individual, and the participants couldn't hear what others were listening at any point during the experiment. The natural light present in the experimental area was sufficient to provide an 'intimate' ambience. Therefore, artificial light was kept to a minimum.

12. Together with the written guidelines concerning the experiment, at least one supervisor was present during the experimental process in order to provide guidance and support. Upon finishing the experiment, the participants were instructed to leave the room without discussing any details

with the next group of participants. The experiment lasted for around 10 minutes.

13. Prior to this ANOVA, the assumption that the covariance matrices of the dependent variables were equal across groups was checked. This was obtained through a Box's Test of Equality of Covariance Matrices. As a measure of effect size, report generalized Partial Eta Squared (η^2) was reported, as suggested by Levine and Hullett (2002).

14. There were no associations between experience preference and music groups [$C = 0.127$, $X^2(2) = 3.616$, $p = 0.164$].

15. Retrieved from https://www.beeradvocate.com/beer/profile/134/349/, April, 2018.

16. Significance at $p < 0.005$.

17. The available choices were TK, WD, or no preference. There were no associations between the group and the preference for these ratings [$C = 0.091$, $X^2(2) = 1.274$, $p = 0.529$].

18. Retrieved from https://www.beeradvocate.com/beer/profile/215/2512/, April 2018.

19. The available choices were TK, WD, or no preference. There were no associations between the group and the preference for these ratings [$C = 0.146$, $X^2(2) = 3.415$, $p = 0.181$].

20. E.g., music that is classified based on the sound characteristics, such as the frequency range, the timbres of the instruments incorporated, and how such sound characteristics can be congruently associated with flavor attributes.

21. https://www.npr.org/sections/thesalt/2014/01/15/262741403/why-sugar-makes-us-feel-so-good, retrieved April 2018.

22. In Experiments 1–3, the participants rated how much they thought each of the music tracks matched the flavor of the beer. No effects were observed in terms of any of these results.

23. https://www.softecspa.com/en/portfolio/campari/, retrieved July, 2018.

24. https://www.moodiedavittreport.com/celebrating-90-years-godiva-hosts-multi-sensory-soiree-in-brussels/, retrieved July, 2018.

25. https://www.moodiedavittreport.com/celebrating-90-years-godiva-hosts-multi-sensory-soiree-in-brussels/, retrieved July, 2018.

26. www.thesoundofchocolate.be, retrieved July, 2018.

27. https://developer.spotify.com/;   https://developer.apple.com/musickit/;
https://developer.amazon.com/; retrieved July, 2018.

References

Aaker, D. A. and Biel, A. L. (2013). *Brand Equity & Advertising: Advertising's Role in Building Strong Brands*. Psychology Press, New York, NY, USA.

Areni, C. S. and Kim, D. (1993). The influence of background music on shopping behavior: classical versus top-forty music in a wine store, *Adv. Consum. Res.* **20**, 336–340.

Arora, M. and Kumar, A. (2018). Consumer awareness towards brand equity, *As. J. Manag.* **9**, 41–53.

Auvray, M. and Spence, C. (2008). The multisensory perception of flavor, *Conscious. Cogn.* **17**, 1016–1031.

Biswas, D., Lund, K. and Szocs, C. (2019). Sounds like a healthy retail atmospheric strategy: effects of ambient music and background noise on food sales, *J. Acad. Mark. Sci.* **47**, 37–55.

Bronner, K., Frieler, K., Bruhn, H., Hirt, R. and Piper, D. (2012). What is the sound of citrus? Research on the correspondences between the perception of sound and flavour, in: *Proceedings of the 12th International Conference of Music Perception and Cognition (ICMPC) and the 8th Triennial Conference of the European Society for the Cognitive Sciences of Music (ESCOM)*. Thessaloniki, Greece, pp. 142–148.

Bruner, G. C. (1990). Music, mood, and marketing, *J. Mark.* **54**, 94–104.

Cheskin, L. (1972). *Marketing Success: How to Achieve It*. Cahners Books, Boston, MA, USA.

Cohen, J. B., Pham, M. T. and Andrade, E. B. (2008). The nature and role of affect in consumer behavior, in: *Handbook of Consumer Psychology*, C. P. Haugtvedt, P. Herr and F. Kardes (Eds), pp. 297–348. Lawrence Erlbaum Associates, New York, NY, USA.

Corr, P. and Plagnol, A. (2018). *Behavioral Economics: the Basics*. Routledge, London, UK.

Cramer, A. O. J., van Ravenzwaaij, D., Matzke, D., Steingroever, H., Wetzels, R., Grasman, R. P. P. P., Waldorp, L. J. and Wagemakers, E.-J. (2016). Hidden multiplicity in exploratory multiway ANOVA: prevalence and remedies, *Psychon. Bull. Rev.* **23**, 640–647.

Crawford, J. R. and Henry, J. D. (2004). The positive and negative affect schedule (PANAS): construct validity, measurement properties and normative data in a large non-clinical sample, *Br. J. Clin. Psychol.* **43**, 245–265.

Crisinel, A.-S. and Spence, C. (2009). Implicit association between basic tastes and pitch, *Neurosci. Lett.* **464**, 39–42.

Crisinel, A.-S. and Spence, C. (2010). As bitter as a trombone: synesthetic correspondences in nonsynesthetes between tastes/flavors and musical notes, *Atten. Percept. Psychophys.* **72**, 1994–2002.

Crisinel, A.-S. and Spence, C. (2012). The impact of pleasantness ratings on crossmodal associations between food samples and musical notes, *Food Qual. Pref.* **24**, 136–140.

Crisinel, A.-S., Cosser, S., King, S., Jones, R., Petrie, J. and Spence, C. (2012). A bittersweet symphony: systematically modulating the taste of food by changing the sonic properties of the soundtrack playing in the background, *Food Qual. Pref.* **24**, 201–204.

Demoulin, N. T. M. (2011). Music congruency in a service setting: the mediating role of emotional and cognitive responses, *J. Retail. Consum. Serv.* **18**, 10–18.

Edwards, A. L. (1957). *The Social Desirability Variable in Personality Assessment and Research*. Dryden Press, Ft Worth, TX, USA.

Fiegel, A., Meullenet, J.-F., Harrington, R. J., Humble, R. and Seo, H.-S. (2014). Background music genre can modulate flavor pleasantness and overall impression of food stimuli, *Appetite* **76**, 144–152.

Fritz, T. H., Brummerloh, B., Urquijo, M., Wegner, K., Reimer, E., Gutekunst, S., Scheinder, L., Smallwood, J. and Villringer, A. (2017). Blame it on the bossa nova: transfer of perceived sexiness from music to touch, *J. Exp. Psychol. Gen.* **146**, 1360–1365.

Hauck, P. and Hecht, H. (2019). Having a drink with Tchaikovsky: the crossmodal influence of background music on the taste of beverages, *Multisens. Res.* **32**(1), 1–24.

Holt-Hansen, K. (1968). Taste and pitch, *Percept. Mot. Skills* **27**, 59–68.

Holt-Hansen, K. (1976). Extraordinary experiences during cross-modal perception, *Percept. Mot. Skills* **43**, 1023–1027.

Huberty, C. J. and Olejnik, S. (2006). *Applied MANOVA and Discriminant Analysis*. John Wiley and Sons, Hoboken, NJ, USA.

Hultén, B. (2011). Sensory marketing: the multi-sensory brand-experience concept, *Eur. Bus. Rev.* **23**, 256–273.

Huron, D. (2007). *Sweet Anticipation: Music and the Psychology of Expectation*. MIT Press, Cambridge, MA, USA.

Johnson, A. and Proctor, R. W. (2004). *Attention: Theory and Practice*. Sage, Thousand Oaks, CA, USA.

Kantono, K., Hamid, N., Shepherd, D., Lin, Y. H. T., Yakuncheva, S., Yoo, M. J. Y., Grazioli, G. and Carr, B. T. (2016b). The influence of auditory and visual stimuli on the pleasantness of chocolate gelati, *Food Qual. Pref.* **53**, 9–18.

Kantono, K., Hamid, N., Shepherd, D., Yoo, M. J. Y., Grazioli, G. and Carr, B. T. (2016a). Listening to music can influence hedonic and sensory perceptions of gelati, *Appetite* **100**, 244–255.

Kantono, K., Hamid, N., Shepherd, D., Yoo, M. J. Y., Carr, B. T. and Grazioli, G. (2016c). The effect of background music on food pleasantness ratings, *Psychol. Music* **44**, 1111–1125.

Kantono, K., Hamid, N., Shepherd, D., Lin, Y. H. T., Brard, C., Grazioli, G. and Carr, B. T. (2018). The effect of music on gelato perception in different eating contexts, *Food Res. Int.* **113**, 43–56.

Knoeferle, K. M., Woods, A., Käppler, F. and Spence, C. (2015). That sounds sweet: using cross-modal correspondences to communicate gustatory attributes, *Psychol. Mark.* **32**, 107–120.

Knöferle, K. and Spence, C. (2012). Crossmodal correspondences between sounds and tastes, *Psychon. Bull. Rev.* **19**, 1–15. DOI:10.3758/s13423-012-0321-z.

Konečni, V. J. (2008). Does music induce emotion? A theoretical and methodological analysis, *Psychol. Aesthet. Creat. Arts* **2**, 115–129.

Krishna, A. (2012). An integrative review of sensory marketing: engaging the senses to affect perception, judgment and behavior, *J. Consum. Psychol.* **22**, 332–351.

Levine, T. R. and Hullett, C. R. (2002). Eta squared, partial eta squared and the misreporting of effect size in communication research, *Hum. Commun. Res.* **28**, 612–625.

Magyar-Moe, J. L. (2009). *Therapist's Guide to Positive Psychological Interventions*. Academic Press, Burlington, MA.

Marin, M. M., Schober, R., Gingras, B. and Leder, H. (2017). Misattribution of musical arousal increases sexual attraction towards opposite-sex faces in females, *PloS One* **12**, e0183531. DOI:10.1371/journal.pone.0183531.

Mattila, A. S. and Wirtz, J. (2001). Congruency of scent and music as a driver of in-store evaluations and behavior, *J. Retail.* **77**, 273–289.

Mesz, B., Trevisan, M. A. and Sigman, M. (2011). The taste of music, *Perception* **40**, 209–219.

Noel, C. and Dando, R. (2015). The effect of emotional state on taste perception, *Appetite* **95**, 89–95.

North, A. C. (2012). The effect of background music on the taste of wine, *Br. J. Psychol.* **103**, 293–301.

North, A. C., Hargreaves, D. J. and McKendrick, J. (1997). In-store music affects product choice, *Nature* **390**, 132.

North, A. C., Hargreaves, D. J. and McKendrick, J. (1999). The influence of in-store music on wine selections, *J. Appl. Psychol.* **84**, 271–276.

Open Science Collaboration (2015). Estimating the reproducibility of psychological science, *Science* **349**, aac4716. DOI:10.1126/science.aac4716.

Piqueras-Fiszman, B. and Spence, C. (Eds) (2016). *Multisensory Flavor Perception: from Fundamental Neuroscience Through to the Marketplace*. Woodhead Publishing, London, UK.

Reinoso Carvalho, F., Van Ee, R., Rychtarikova, M., Touhafi, A., Steenhaut, K., Persoone, D., Spence, C. and Leman, M. (2015a). Does music influence the multisensory tasting experience?, *J. Sens. Stud.* **30**, 404–412.

Reinoso Carvalho, F., Van Ee, R., Touhafi, A., Steenhaut, K. and Rychtarikova, M. (2015b). Assessing multisensory tasting experiences by means of customized soundscapes, in: *Euronoise 2015, Maastricht, The Netherlands, Vol. 1*, pp. 739–744.

Reinoso Carvalho, F., Van Ee, R., Rychtarikova, M., Touhafi, A., Steenhaut, K., Persoone, D. and Spence, C. (2015c). Using sound–taste correspondences to enhance the subjective value of tasting experiences, *Front. Psychol.* **6**, 1309. DOI:10.3389/fpsyg.2015.01309.

Reinoso Carvalho, F., Wang, Q. (J.), de Causmaecker, B., Steenhaut, K., van Ee, R. and Spence, C. (2016a). Tune that beer! Listening for the pitch of beer, *Beverages* **2**, 31. DOI:10.3390/beverages2040031.

Reinoso Carvalho, F., Steenhaut, K., van Ee, R., Touhafi, A. and Velasco, C. (2016b). Sound-enhanced gustatory experiences and technology, in: *Proceedings of the 1st Workshop on Multi-Sensorial Approaches to Human-Food Interaction, Tokyo, Japan*, art. 5.

Reinoso Carvalho, F., Velasco, C., van Ee, R., Leboeuf, Y. and Spence, C. (2016c). Music influences hedonic and taste ratings in beer, *Front. Psychol.* **7**, 636. DOI:10.3389/fpsyg.2016.00636.

Reinoso Carvalho, F., Wang, Q. J., Van Ee, R. and Spence, C. (2016d). The influence of soundscapes on the perception and evaluation of beers, *Food Qual. Pref.* **52**, 32–41.

Reinoso Carvalho, F., Wang, Q. (J.), van Ee, R., Persoone, D. and Spence, C. (2017). "Smooth operator": music modulates the perceived creaminess, sweetness, and bitterness of chocolate, *Appetite* **108**, 383–390.

Rubin, M. (2016). The perceived awareness of the research hypothesis scale: assessing the influence of demand characteristics, *Figshare*. DOI:10.6084/m9.figshare.4315778.

Rudmin, F. and Cappelli, M. (1983). Tone–taste synesthesia: a replication, *Percept. Mot. Skills* **56**, 118.

Schwarz, N. and Clore, G. L. (1983). Mood, misattribution, and judgments of well-being: informative and directive functions of affective states, *J. Personal. Soc. Psychol.* **45**, 513–523.

Sester, C., Deroy, O., Sutan, A., Galia, F., Desmarchelier, J.-F., Valentin, D. and Dacremont, C. (2013). "Having a drink in a bar": an immersive approach to explore the effects of context on drink choice, *Food Qual. Pref.* **28**, 23–31.

Shaffer, J. P. (1995). Multiple hypothesis testing, *Annu. Rev. Psychol.* **46**, 561–584.

Spence, C. (2011). Crossmodal correspondences: a tutorial review, *Atten. Percept. Psychophys.* **73**, 971–995.

Spence, C. (2012). Auditory contributions to flavour perception and feeding behaviour, *Physiol. Behav.* **107**, 505–515.

Spence, C. (2014). Noise and its impact on the perception of food and drink, *Flavour* **3**, 9. DOI:10.1186/2044-7248-3-9.

Spence, C. (2017a). *Gastrophysics: the New Science of Eating.* Viking Penguin, London, UK.

Spence, C. (2017b). Tasting in the air: a review, *Int. J. Gastron. Food Sci.* **9**, 10–15.

Spence, C. (2017c). Sonic seasoning, in: *Audio Branding: Using Sound to Build Your Brand*, pp. 52–58, L. Minsky and C. Fahey (Eds). Kogan Page, London, UK.

Spence, C. (2019a). Multisensory experiential wine marketing, *Food Qual. Pref.* **71**, 106–116.

Spence, C. (2019b). On the relative nature of (pitch-based) crossmodal correspondences, *Multisens. Res.* **32**, 235–265. DOI:10.1163/22134808-20191407.

Spence, C., Shankar, M. U. and Blumenthal, H. (2011). 'Sound bites': auditory contributions to the perception and consumption of food and drink, in: *Art and the Senses*, pp. 207–238, F. Bacci and D. Melcher (Eds). Oxford University Press, New York, NY, USA.

Spence, C., Richards, L., Kjellin, E., Huhnt, A.-M., Daskal, V., Scheybeler, A., Velasco, C. and Deroy, O. (2013). Looking for crossmodal correspondences between classical music and fine wine, *Flavour* **2**, 29. DOI:10.1186/2044-7248-2-29.

Spence, C., Michel, C. and Smith, B. (2014). Airplane noise and the taste of umami, *Flavour* **3**, 2. DOI:10.1186/044-7248-3-2.

Velasco, C., Carvalho, F. R., Petit, O. and Nijholt, A. (2016). A multisensory approach for the design of food and drink enhancing sonic systems, in: *Proceedings of the 1st Workshop on Multi-Sensorial Approaches to Human-Food Interaction, Tokyo, Japan*, p. 7.

Velasco, C., Jones, R., King, S. and Spence, C. (2013). Assessing the influence of the multi-sensory environment on the whisky drinking experience, *Flavour* **2**, 23. DOI:10.1186/2044-7248-2-23.

Wang, Q. (J.) and Spence, C. (2015a). Assessing the effect of musical congruency on wine tasting in a live performance setting, *i-Perception* **6**, 2041669515593027. DOI:10.1177/2041669515593027.

Wang, Q. J. and Spence, C. (2015b). Assessing the influence of the multisensory atmosphere on the taste of vodka, *Beverages* **1**, 204–217.

Wang, Q. (J.) and Spence, C. (2016). "Striking a sour note": assessing the influence of consonant and dissonant music on taste perception, *Multisens. Res.* **29**, 195–208.

Wang, Q. (J.) and Spence, C. (2017). Assessing the role of emotional associations in mediating crossmodal correspondences between classical music and red wine, *Beverages* **3**, 1. DOI:10.3390/beverages3010001.

Wang, Q. (J.) and Spence, C. (2018). "A sweet smile": the modulatory role of emotion in how extrinsic factors influence taste evaluation, *Cogn. Emot.* **32**, 1052–1061.

Watson, D., Clark, L. A. and Tellegen, A. (1988). Development and validation of brief measures of positive and negative affect: the PANAS scales, *J. Personal. Soc. Psychol.* **54**, 1063–1070.

Watson, Q. J. and Gunter, K. L. (2017). Trombones elicit bitter more strongly than do clarinets: a partial replication of three studies of Crisinel and Spence, *Multisens. Res.* **30**, 321–335.

Woods, A. T., Poliakoff, E., Lloyd, D. M., Dijksterhuis, G. B. and Thomas, A. (2010). Flavor expectation: the effects of assuming homogeneity on drink perception, *Chemosens. Percept.* **3**, 174–181.

Yuan, Z., Ghinea, G. and Muntean, G.-M. (2015). Beyond multimedia adaptation: quality of experience-aware multi-sensorial media delivery, *IEEE Trans. Multimedia* **17**, 104–117.

Zellner, D., Geller, T., Lyons, S., Pyper, A. and Riaz, K. (2017). Ethnic congruence of music and food affects food selection but not liking, *Food Qual. Pref.* **56**, 126–129.

A Sweet Voice: The Influence of Cross-Modal Correspondences Between Taste and Vocal Pitch on Advertising Effectiveness

Kosuke Motoki [1,2,3,*], **Toshiki Saito** [2,3], **Rui Nouchi** [2], **Ryuta Kawashima** [2] and **Motoaki Sugiura** [2,4]

[1] Department of Food Management, Miyagi University, Sendai, Japan
[2] Institute of Development, Aging and Cancer, Tohoku University, Sendai, Japan
[3] Japan Society for the Promotion of Science, Tokyo, Japan
[4] International Research Institute of Disaster Science, Tohoku University, Sendai, Japan

Abstract

We have seen a rapid growth of interest in cross-modal correspondences between sound and taste over recent years. People consistently associate higher-pitched sounds with sweet/sour foods, while lower-pitched sounds tend to be associated with bitter foods. The human voice is key in broadcast advertising, and the role of voice in communication generally is partly characterized by acoustic parameters of pitch. However, it remains unknown whether voice pitch and taste interactively influence consumer behavior. Since consumers prefer congruent sensory information, it is plausible that voice pitch and taste interactively influence consumers' responses to advertising stimuli. Based on the cross-modal correspondence phenomenon, this study aimed to elucidate the role played by voice pitch–taste correspondences in advertising effectiveness. Participants listened to voiceover advertisements (at a higher or lower pitch than the original narrator's voice) for three food products with distinct tastes (sweet, sour, and bitter) and rated their buying intention (an indicator of advertising effectiveness). The results show that the participants were likely to exhibit greater buying intention toward both sweet and sour food when they listened to higher-pitched (vs lower-pitched) voiceover advertisements. The influence of a higher pitch on sweet and sour food preferences was observed in only two of the three studies: studies 1 and 2 for sour food, and studies 2 and 3 for sweet food. These findings emphasize the role that voice pitch–taste correspondence plays in preference formation, and advance the applicability of cross-modal correspondences to business.

Keywords

Cross-modal correspondences, high pitch, tastes, voice, advertising

* To whom correspondence should be addressed. E-mail: motokik@myu.ac.jp

© KONINKLIJKE BRILL NV, LEIDEN, 2019 | DOI:10.1163/9789004416307_008

1. Introduction

People often map sensory information onto the other senses in a surprisingly consistent manner. Cross-modal correspondences refer to the tendency for humans (and non-human species) to preferentially associate certain features or dimensions of stimuli across different sensory modalities (see Spence, 2011 for a review). A variety of cross-modal correspondences across the senses have been reported, including associations between taste and sound (e.g., Crisinel *et al.*, 2012a; Knöferle and Spence, 2012; Knoeferle *et al.*, 2015; Simner, Cuskley and Kirby, 2010), taste and shapes (e.g., Motoki *et al.*, 2019a; Ngo and Spence, 2011; Ngo *et al.*, 2011; Velasco *et al.*, 2016), sound and shapes (e.g., Bremner *et al.*, 2013; Knoeferle *et al.*, 2017; Ramachandran and Hubbard, 2001; Spence, 2012), odor and sounds (e.g., Crisinel and Spence, 2011; Deroy *et al.*, 2013), and warmth and color (e.g., Ho *et al.*, 2014; Motoki *et al.*, 2019b). For example, sweet tastes are consistently associated with high-pitched sounds (Crisinel and Spence, 2009), piano instruments (Knöferle and Spence, 2012), and round shapes (Velasco *et al.*, 2016), while bitter tastes are paired with low pitched-sounds (Crisinel and Spence, 2009; Wang *et al.*, 2015), brass instruments (Knöferle and Spence, 2012), and angular shapes (Velasco *et al.*, 2016).

The cross-modal correspondence phenomenon has recently been applied to the study of consumer behavior. Store atmospherics affect buying behavior, so manipulating the genre or volume of music in stores influences food and beverage purchases (e.g., Biswas *et al.*, 2019; North *et al.*, 1999). For example, low volume music or noise increases unhealthy food purchases (Biswas *et al.*, 2019). Importantly, store atmospherics are fundamentally multisensory, and consumers experience the simultaneous stimulation of multiple senses (Spence *et al.*, 2014). In view of the multisensory nature of consumer experience, previous studies have investigated how multisensory interactions influence consumer behavior; warm sensations have been observed to guide visual attention and preference toward light-colored (vs dark-colored) products (Motoki *et al.*, 2019b). Low-frequency (vs high-frequency) sounds guide visual attention and preference toward dark-colored (vs light-colored) products (Hagtvedt and Brasel, 2016). Products occupying upper (vs lower) shelves are more desirable when they are sweet (Velasco *et al.*, 2019) or light-colored (Sunaga *et al.*, 2016). These findings all indicate that congruent sensory information positively influences consumer preference.

The human voice plays an important role in marketing communication, and the role of voice in communication generally is partly characterized by the acoustic parameter of pitch. Voices are ubiquitous in broadcast advertising (i.e., TV, radio, or internet commercials): actors or actresses regularly appear in TV commercials and convey marketing messages using their voices. Previous studies have investigated the effects of voice on consumer behavior (e.g., Chattopadhyay *et al.*, 2003; Chebat *et al.*, 2007; Martín-Santana *et al.*, 2015). Findings indicate that, for example, lower-pitched voices generate a more positive attitude toward the spokesperson (Martín-Santana *et al.*, 2015) and the product being advertised (Chattopadhyay *et al.*, 2003). Although evidence suggests that vocal acoustic features are likely to affect advertising effectiveness, the mechanisms of the interaction between voice and product type that influence consumer responses to advertising remain largely uninvestigated. Sensory variables (e.g., pitch) do not always act in isolation to influence consumer behavior; multiple sensory variables can interactively exert an influence on consumers (Hagtvedt and Brasel, 2016; Motoki *et al.*, 2018, 2019b; Sunaga *et al.*, 2016; Velasco *et al.*, 2019). Considering the copious amounts of money that are consistently invested in broadcast advertising (Emarketers, 2018a, b), it is important to understand in detail how the acoustic features of voice (e.g., pitch) influence advertising effectiveness.

Based on the cross-modal correspondence phenomenon, the present study focused on vocal pitch, and investigated how it interacts with the other senses relevant to product attributes to influence consumer behavior. The cross-modal correspondence phenomenon offers a novel starting hypothesis: sour tastes are consistently associated with high-pitched sounds, while bitter tastes are paired with low-pitched sounds (e.g., Crisinel and Spence, 2009). Additionally, high pitch is generally associated with sweet taste (see Knöferle and Spence, 2012 for a review), though there is some inconsistency in the findings related to this (i.e., lower pitch is occasionally associated with sweet taste) (Velasco *et al.*, 2014). Previous studies have verified that sounds can modify tastes (Crisinel *et al.*, 2012b; Wang and Spence, 2016; Wang *et al.*, 2017; Watson and Gunther, 2017). For example, a soundtrack designed to modify sweetness (vs bitter or sour soundtracks) increased consumers' perceptions of the sweetness and their enjoyment of beer (Reinoso Carvalho *et al.*, 2016). Significantly, vocal congruity in advertising contributes to communication effectiveness by enhancing purchase intent and brand attitude (Oakes, 2007). Considering that consumers tend to prefer congruent sensory information (e.g., Hagtvedt and Brasel, 2016; Motoki *et al.*, 2019b; Sunaga *et al.*, 2016; Velasco *et al.*, 2019), it is plausible that vocal pitch and taste interactively guide consumer response to advertising. Congruent sensory information (i.e., sweet/sour taste and high pitch, bitter taste and low pitch) may elicit positive responses to advertisements. However, whether correspondences between voice and taste influence advertising effectiveness remains unconfirmed.

This study aimed to elucidate the mechanisms by which vocal acoustic features and product types interactively influence consumer responses to advertisements. Based on the cross-modal correspondence phenomenon, we investigated whether voice pitch (higher/lower) and taste perceptions (sweet/sour/bitter) interactively influence advertising effectiveness. The participants listened to broadcast advertisements for sweet, sour, and bitter foods. Higher or lower voiceover pitches were randomly allocated to the participants. Based on the pitch–taste correspondence phenomenon, we may predict that congruent sensory advertisements are more appealing. Especially, we hypothesize that advertisements for sweet/sour (bitter) foods are more appealing when the narrator's voice is higher (lower).

2. Study 1

2.1. Method

2.1.1. Design
Our study examined whether voiceover messages with different pitches would influence buying intention with regard to foods that had distinct perceived tastes. The study had a 2 (pitch: higher, lower) × 3 (perceived tastes: sweet, sour, bitter) mixed design, wherein pitch represented the between-participants and taste the within-participant factor. The main outcome was buying intention with regard to food products.

2.1.2. Participants
In total, 59 healthy participants (28 females; $M_{age} = 21.12 \pm 1.96$ years) were recruited via bulletin-board postings and a student mailing list email. The sample size was determined before data collection was initiated based on the suggestion that there be at least 20 participants per cell (Simmons *et al.*, 2011) (however, see the revised suggestions for sample size [Simmons *et al.*, 2018] that were brought to our attention after the data collection was completed). A between-participants design confers benefits by reducing the likelihood that the participants would infer the research aim. If a within-participant design had been used, the participants would have listened to both higher and lower vocal pitches and may have deduced the study aim (e.g., to determine relationships between vocal pitch and distinct tastes) and may have tended to provide socially acceptable responses (i.e., those apparently favored by the experimenters). Additionally, given the widespread indication that pitch-based cross-modal correspondences are more relative than absolute, the between-participants design may minimize the relative effects of cross-modal correspondences (Spence, 2019). Thus, we used a between-subjects design to eliminate this problem. The study was approved by the ethics com-

mittee of the School of Medicine at Tohoku University and was conducted in accordance with the Declaration of Helsinki.

2.1.3. Task

Participants were randomly allocated to higher-pitch ($n = 30$, 14 females) and lower-pitch ($n = 29$, 14 females) groups. Voice pitches were manipulated using Audacity® (https://www.audacityteam.org/). We used the *Change Pitch* implemented in Audacity (accessed by *Effect > Change Pitch*) to change the vocal pitch in the advertisement. The manipulation was conducted on the basis of previous studies that had used Audacity to change human vocal pitch (Puts *et al.*, 2006; Stel *et al.*, 2012). The higher-pitch condition was achieved by increasing the pitch frequency by 20% and 3.16 semitones, while the lower-pitch condition was achieved by decreasing the pitch frequency by 30% and -6.17 semitones.

A professional female narrator from the Two-eight Company (www.28inc. co.jp/) recorded the voiceovers for the food advertisements (sweet: cream puff, sour: lemon juice, bitter: black coffee). The advertising messages were designed by the experimental team so as to be as homogeneous as possible with regard to duration and content. The voiceover script for the sweet food was "Enjoy a creamy texture and sweet vanilla flavor"; the script for the sour food was "Enjoy a refreshingly sour and concentrated lemon flavor" and the script for the bitter food was "Enjoy a clear and full-bodied bitter taste".

Groups of two to four participants were placed in a room that accommodated a maximum of 10 people, where they listened, without delay, to the advertisements for each of the three foods. The voiceover messages were aired using a MacBook's built-in speakers. The volume was the same for all advertisements (with the MacBook's volume control at its highest), and all messages had almost the same duration (8 s for the sweet food, 7 s for the sour food, and 8 s for the bitter food). The order in which each food advertisement was presented was randomized across the groups. Having listened to each voiceover message, the participants were asked to rate their buying intention and their taste perceptions (e.g., "How much do you want to buy this product?", "How sweet is this?", "How sour is this?", "How bitter is this?"). The participants' ratings in response to each question were recorded using a seven-point scale that ranged from 1 = "Not at all" to 7 = "Very much". Finally, the participants were asked to answer two additional questions: "How much do you like this voice?" and "How high-pitched is this voiceover advertisement?" Respondents again provided their ratings using a seven-point scale ranging from 1 = "Not at all" to 7 = "Very much" for the first question and from 1 = "Very low" to 7 = "Very high" for the second question.

For each voiceover, the fundamental frequency F0, F1, F2, and spectral balance were calculated using VoiceSauce (Shue *et al.*, 2011). F0 was calculated using the Straight algorithm (Shue *et al.*, 2011). F1 and F2 were calculated using the Snack algorithm (Shue *et al.*, 2011). Spectral balance was calculated using the R package "TuneR" (Ligges *et al.*, 2018). These parameters are shown in Appendix Table A1.

2.2. Statistical Analysis

A *t*-test was performed to determine whether the participants perceived pitch differently and expressed different preferences in response to the advertisements, according to whether the conditions were high- or low-pitched. Furthermore, we conducted an ANOVA to assess the effects of higher- vs lower-pitched voices on taste perception (sweetness, sourness, bitterness), as well as buying intention. The design format was 2 (voice pitch: high, low) × 3 (tastes: sweet, sour, bitter), in which voice pitch functioned as the between-participants factor and taste as the within-participant factor. The dependent variables were the taste perception ratings and buying intention. Where a significant interaction was identified, we conducted post-hoc analyses to achieve a fuller understanding of the interaction's details. The post-hoc analyses were conducted using Shaffer's modified sequentially rejective Bonferroni test procedure. Where sphericity was violated, we applied the Huynh–Feldt correction.

2.3. Results

2.3.1. Pitch Perception
We tested to determine whether the participants perceived the vocal pitch in the advertising messages differently depending on whether they listened to them at the high- or low-pitched settings. The participants in the higher-pitch conditions perceived the pitch as higher than did those who listened in the lower-pitch conditions ($M_{higher} = 5.13 \pm 1.04$ vs $M_{ower} = 4.17 \pm 1.49$, Cohen's $d = 0.750$; $t_{57} = 2.879$, $p = 0.006$, Cohen's $d = 0.098$). The preference with regard to pitch did not differ between the higher- and lower-pitch conditions ($M_{higher} = 3.83 \pm 1.34$ vs $M_{lower} = 3.69 \pm 1.58$; $t_{57} = 0.377$, $p = 0.708$). These findings confirmed that pitch perception had been successfully manipulated, while the preferences with regard to higher and lower pitch were controlled.

2.3.2. Sweet Perception
We tested to determine whether the participants perceived the sweet food as sweeter than the other foods. We used sweetness as the dependent variable in an ANOVA, in which pitch (higher, lower) functioned as the between-participants factor, with perceived taste (sweet, sour, bitter) as the within-participant factor.

No main effect was associated with the vocal pitch in the advertising message [$F(1, 57) = 2.314$, $p = 0.134$, $\eta_p^2 = 0.039$] nor with the interaction between pitch and taste [$F(2, 114) = 0.797$, $p = 0.453$, $\eta_p^2 = 0.014$]. However, the main effect of taste was significant [$F(2, 114) = 493.325$, $p < 0.001$, $\eta_p^2 = 0.896$], indicating that participants perceived food sweetness differently, depending on the food's taste. Post-hoc analyses showed that the participants' perception of sweetness was more heightened for sweet food ($M = 5.85 \pm 0.93$) than for sour ($M = 1.83 \pm 1.00$: $t_{57} = 23.286$, adj.$p < 0.001$) and bitter food ($M = 1.39 \pm 0.62$: $t_{57} = 29.202$, adj.$p < 0.001$). Participants also exhibited a heightened perception of sweetness with regard to sour food, to a greater extent than for bitter food ($t_{57} = 3.069$, adj.$p = 0.003$).

2.3.3. Sour Perception

We tested to determine whether the participants perceived the sour food as sourer than the other foods. Sourness was used as the dependent variable in an ANOVA, in which pitch (higher, lower) functioned as the between-participants factor and perceived taste (sweet, sour, bitter) as the within-participant factor.

No main effect was associated with vocal pitch in the advertising messages [$F(1, 57) = 0.871$, $p = 0.355$, $\eta_p^2 = 0.015$], but the main effect of taste was significant [$F(2, 114) = 428.610$, $p < 0.001$, $\eta_p^2 = 0.883$], indicating that participants perceived sourness differently depending on the food's taste. Post-hoc analyses showed that the participants' perception of sourness was heightened with regard to sour food, ($M = 6.26 \pm 0.80$) to a greater extent than for sweet food ($M = 1.24 \pm 0.75$: $t_{57} = 38.247$, adj.$p < 0.001$) or for bitter food ($M = 2.43 \pm 1.48$: $t_{57} = 19.856$, adj.$p < 0.001$). Participants also exhibited a greater perception of sourness with regard to sour food than to bitter food ($t_{57} = 5.814$, adj.$p < 0.001$).

Notably, there was a significant interaction between pitch and sourness [$F(2, 114) = 2.449$, $p = 0.091$, $\eta_p^2 = 0.041$]. A planned comparison revealed that higher-pitched voice advertising (vs lower-pitched) decreased sourness perception in relation to sour food [higher: 6.00 ± 0.83 vs lower: 6.52 ± 0.69, $F(1, 57) = 6.765$, $p = 0.012$, $\eta_p^2 = 0.106$]. By contrast, the advertising message, when communicated in a higher (vs lower) pitch did not influence perceived sourness with regard to sweet food [higher: $M = 1.37 \pm 0.89$ vs lower: $M = 1.10 \pm 0.56$, $F(1, 57) = 1.840$, $p = 0.180$, $\eta_p^2 = 0.031$] or bitter food [higher: $M = 2.30 \pm 1.44$ vs lower: $M = 2.55 \pm 1.53$, $F(1, 57) = 0.425$, $p = 0.517$, $\eta_p^2 = 0.007$].

2.3.4. Bitter Perception

We tested to determine whether the participants perceived the bitter food as more bitter than the other foods. We used bitterness as the dependent variable in an ANOVA, in which pitch (higher, lower) functioned as the between-participants factor and taste perception (sweet, sour, bitter) as the within-participants factor.

There was no main effect associated with the vocal pitch of the advertising voiceover [$F(1, 57) = 0.037$, $p = 0.849$, $\eta_p^2 < 0.001$] or with the interaction between pitch and taste [$F(2, 114) = 0.564$, $p = 0.571$, $\eta_p^2 = 0.010$]. However, the main effect of taste was significant [$F(2, 114) = 431.210$, $p < 0.001$, $\eta_p^2 = 0.883$], indicating that participants perceived bitterness differently depending on the food tastes. Post-hoc analyses verified that the participants' perception of bitterness was more heightened for bitter food ($M = 5.73 \pm 0.76$) than for sweet food ($M = 1.17 \pm 0.77$: $t_{57} = 34.552$, adj.$p < 0.001$) or sour food ($M = 1.95 \pm 1.40$: $t_{57} = 19.157$, adj.$p < 0.001$). Participants' perception of bitterness was also more heightened for sour food than for sweet food ($t_{57} = 4.812$, adj.$p = 0.003$).

2.3.5. Buying Intention

We carried out experiments to determine whether the participants' food preferences were affected by vocal pitch–taste correspondences. We used buying intention as the dependent variable in an ANOVA, in which pitch (higher, lower) functioned as the between-participants factor, with perceived taste (sweet, sour, bitter) as the within-participant factor.

There was no main effect associated with the vocal pitch of the voiceover advertisements [$F(1, 57) < 0.001$, $p = 0.982$, $\eta_p^2 < 0.001$]. However, the main effect of taste was significant [$F(2, 57) = 4.544$, $p = 0.013$, $\eta_p^2 = 0.074$, indicating that participants prefer foods differentially as a function of the food's taste. Post-hoc analyses indicated that the participants exhibited greater buying intention toward sweet food ($M = 4.58 \pm 1.38$) than toward sour food ($M = 4.03 \pm 1.45$: $t_{57} = 2.384$, $p = 0.021$, adj.$p = 0.035$) or bitter food ($M = 3.92 \pm 1.52$: $t_{57} = 2.603$, $p = 0.012$, adj.$p = 0.035$). Sour and bitter foods did not differ in terms of their associated buying intention ratings ($t_{57} = 0.496$, $p = 0.622$, adj.$p = 0.622$).

Notably, and consistent with our prediction, there was a significant interaction between vocal pitch and taste [$F(2, 57) = 5.308$, $p = 0.006$, $\eta_p^2 = 0.085$]. A planned comparison revealed that the higher-pitched (vs lower-pitched) voiceover message increased buying intention toward sour food [higher: $M = 4.47 \pm 1.196$ vs lower: $M = 3.59 \pm 1.570$, $F(1, 57) = 5.897$, $p = 0.018$, $\eta_p^2 = 0.094$]. By contrast, the higher-pitched (vs lower-pitched) voiceover advertisement did not influence buying intention toward sweet food [higher: $M = 4.30 \pm 1.47$ vs lower: $M = 4.86 \pm 1.25$, $F(1, 57) = 2.511$, $p = 0.119$, $\eta_p^2 = 0.042$] or bitter food [higher: $M = 3.77 \pm 1.50$ vs lower: $M = 4.07 \pm 1.56$, $F(1, 57) = 0.577$, $p = 0.451$, $\eta_p^2 = 0.010$]. The results are presented in Fig. 1.

2.3.6. Discussion

We have shown that higher- (vs lower-) pitched voiceover advertisements increased buying intention toward sour food. In contrast, we did not find any

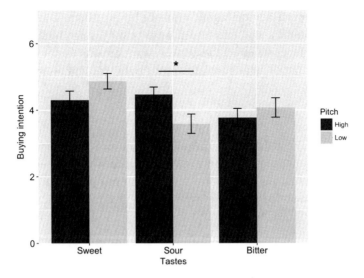

Figure 1. Results of Study 1. Interactive effects of taste and pitch on buying intention. Advertisements in a higher-pitched (vs lower-pitched) voice increased buying intention toward sour food. By contrast, advertisements in a higher-pitched (vs lower-pitched) voice did not influence buying intention toward sweet or bitter foods. The buying intention ratings were rated on a seven-point scale, with 1 = not at all and 7 = very much. Error bars show standard error. Asterisks indicate statistical significance (*, $p < 0.05$).

evidence that changes in the pitch of voiceover advertisements influenced buying intention toward sweet or bitter food.

One concern with this study was that the contents of the message used ("Enjoy a refreshingly sour and concentrated lemon flavor," etc.) might have affected the influence of pitch on buying intention toward food. The other concern was that the level of pitch change was inconsistent between the higher and lower conditions. The higher pitch was increased 20% from the original women's voice, while the lower pitch was decreased 30% from the original women's voice. In a follow-up study, we aimed to (1) replicate the pitch effects of sour food and (2) identify the effects of pitch on foods using a much higher pitched-voiceover advertisement with a simplified message and the same degree of change in pitch (25% pitch increase for higher pitch, 25% pitch decrease for lower pitch).

3. Study 2

3.1. Method

3.1.1. Design
Using simplified voiceover advertisements with much higher pitch, the second study aimed to (1) replicate the effects of higher pitch on sour food,

and (2) clarify the effects of higher pitch on sweet food. The study had a 2 (pitch: higher, lower) × 3 (perceived taste: sweet, sour, bitter) mixed design, wherein pitch represented the between-participants factor and taste the within-participant factor. The main outcome was buying intention toward food products.

3.1.2. Participants

In total, 100 healthy participants (33 females; $M_{age} = 41.30$ years, SD = 9.00) were recruited through Lancers (https://www.lancers.jp/) and completed the study survey on Qualtrics (https://www.qualtrics.com/jp/). As a study prerequisite, we asked participants to confirm that they had an audio-capable device and headphones for the study.

3.1.3. Task Procedure

As in Study 1, voice pitch was manipulated using Audacity® (https://www. audacityteam.org/), but in this study, pitch was adjusted 25% up or down for the higher pitch and lower pitch conditions, respectively. The same professional female narrator as in Study 1 recorded the voiceovers for the food advertisements (sweet: cream puff; sour: lemon juice; bitter: black coffee). The voiceover script was simplified. The script for the sweet food was "Would you like a cream puff?", the script for the sour food was "Would you like a lemon juice?" and the script for the bitter food was "Would you like a black coffee?" As in Study 1, we calculated F0, F1, F2, and spectral balance, and these values are shown in Appendix Table B.

We first asked participants to put on their headphones and confirm that they could hear the sound file that was playing. Participants were then randomly allocated to high pitch ($n = 48$, 13 females) and low pitch ($n = 52$, 20 females) groups. As in Study 1, the participants, after having listened to each voiceover message, were asked to indicate their buying intention and taste perceptions (e.g., "How much do you want to buy this product?", "How sweet is this?", "How sour is this?", "How bitter is this?"). Finally, the participants listened to the sample voiceover script ("Would you like this product?") and were asked to rate their perception of sweet/sour/bitter ("How sweet is this?"/"How sour is this?"/"How bitter is this?") on a 7-point Likert scale. Then they answered the following question about pitch: "How high-pitched is the female voiceover in this advertisement?: 1 = Not at all for a woman to 7 = Very much for a woman." They also rated the voiceover's attributes on a 7-point Likert scale, including valence ("very negative" to "very positive"), arousal, femininity/masculinity, pitch, and volume ("not at all" to "very much" for each attribute).

3.2. Results

3.2.1. Pitch Perception

We assessed whether the participants in this study perceived the vocal pitch of advertising messages differently depending on whether they listened to them under higher- or lower-pitched conditions. The participants in the higher pitch condition perceived the pitch as higher than did those in the lower pitch condition ($M_{higher} = 4.50 \pm 1.19$ vs $M_{lower} = 2.42 \pm 1.18$; $t_{98} = 8.786$, $p < 0.001$, Cohen's $d = 1.759$). The participants in the higher pitch condition perceived the voice to be less masculine (i.e., more feminine) than did those in the low pitch condition ($M_{higher} = 2.29 \pm 0.80$ vs $M_{lower} = 4.27 \pm 1.48$, Cohen's $d = -1.642$; $t_{98} = -8.203$, $p < 0.001$). Perceptions of volume ($M_{higher} = 4.19 \pm 0.73$ vs $M_{lower} = 4.29 \pm 0.70$; $t_{98} = -0.706$, $p = 0.482$), valence (preference; $M_{higher} = 4.45 \pm 0.88$ vs $M_{lower} = 4.19 \pm 1.07$; $t_{97} = 1.287$, $p = 0.201$), and arousal ($M_{higher} = 3.11 \pm 1.34$ vs $M_{lower} = 3.04 \pm 1.33$; $t_{97} = 0.253$, $p = 0.801$) did not differ between the higher and lower pitch conditions. Sweet matching ($M_{higher} = 3.99 \pm 1.30$ vs $M_{lower} = 3.85 \pm 1.42$; $t_{148} = 0.622$, $p = 0.535$), sour matching ($M_{higher} = 3.04 \pm 1.30$ vs $M_{lower} = 2.62 \pm 1.19$; $t_{148} = 1.963$, $p = 0.052$), and bitter matching ($M_{higher} = 2.87 \pm 1.34$ vs $M_{lower} = 2.42 \pm 1.24$; $t_{148} = 1.985$, $p = 0.049$) did not differ between the higher and lower pitch conditions. These findings confirmed that pitch perception had been successfully manipulated, while preferences with regard to higher and lower pitch were controlled.

3.2.2. Perception of Sweetness

We tested whether the participants perceived the sweet food as sweeter than the other foods. We used sweetness as the dependent variable in an ANOVA, in which pitch (higher, lower) functioned as the between-participants factor and taste (sweet, sour, bitter) as the within-participant factor.

There was no main effect associated of the vocal pitch of voiceover advertisements [$F(1, 98) = 1.357$, $p = 0.247$, $\eta_p^2 = 0.014$]. However, the main effect of taste was significant [$F(1.81, 177.04) = 60.973$, $p < 0.001$, $\eta_p^2 = 0.384$], indicating that participants perceived foods differentially depending on the taste. Post-hoc analyses showed that the participants' perception of sweetness was greater for sweet food ($M = 4.59 \pm 1.31$) than for sour ($M = 3.07 \pm 1.29$: $t_{98} = 9.301$, adjusted $p < 0.001$) and bitter foods ($M = 3.01 \pm 1.59$: $t_{98} = 8.539$, adjusted $p < 0.001$). The perception of sweetness did not differ between sour and bitter food ($t_{98} = 0.400$, adjusted $p = 0.690$). There was no significant interaction between vocal pitch and taste [$F(1.81, 177.04) = 0.430$, $p = 0.631$, $\eta_p^2 = 0.004$].

3.2.3. Perception of Sourness

We tested whether the participants perceived the sour food as more sour than the other foods. We used sourness as the dependent variable in an ANOVA, in which pitch (higher, lower) functioned as the between-participants factor and taste (sweet, sour, bitter) as the within-participant factor.

There was no main effect of the vocal pitch of voiceover advertisements $[F(1, 98) = 0.152, p = 0.697, \eta_p^2 = 0.002]$. However, the main effect of taste was significant $[F(1.77, 173.01) = 66.555, p < 0.001, \eta_p^2 = 0.405]$, indicating that participants perceived foods differently depending on the food taste. Post-hoc analyses showed that the participants' perception of sourness was greater for sour food ($M = 4.22 \pm 1.55$) than for sweet ($M = 2.22 \pm 1.31$: $t_{98} = 9.608$, adjusted $p < 0.001$) and bitter foods ($M = 2.86 \pm 1.31$: $t_{98} = 8.227$, adjusted $p < 0.001$). The perception of sourness was also greater for the bitter food than for sweet food ($t_{98} = 4.190$, adjusted $p < 0.001$). There was no significant interaction between vocal pitch and taste $[F(1.77, 173.01) = 0.014, p = 0.979, \eta_p^2 = 0.0001]$.

3.2.4. Perception of Bitterness
We tested whether the participants perceived the bitter food as more bitter than the other foods. We used bitterness as the dependent variable in an ANOVA, in which pitch (higher, lower) functioned as the between-participants factor and taste (sweet, sour, bitter) as the within-participant factor.

There was no main effect of the vocal pitch of voiceover advertisements $[F(1, 98) = 1.388, p = 0.242, \eta_p^2 = 0.014]$. However, the main effect of taste was significant $[F(1.72, 168.14) = 66.455, p < 0.001, \eta_p^2 = 0.404]$, indicating that participants perceived foods differently depending on the food taste. Post-hoc analyses showed that the participants' perception of bitterness was greater for bitter food ($M = 3.91 \pm 1.49$) than for sweet ($M = 2.05 \pm 1.21$: $t_{98} = 10.348$, adjusted $p < 0.001$) and sour foods ($M = 2.64 \pm 1.32$: $t_{98} = 6.919$, adjusted $p < 0.001$). The perception of bitterness was also greater for sour food than for sweet food ($t_{98} = 4.729$, adjusted $p < 0.001$). There was no significant interaction between vocal pitch and taste $[F(1.72, 168.14) = 0.001, p = 0.998, \eta_p^2 = 0.000]$.

3.2.5. Buying Intention
We tested to determine whether the participants' food preferences were affected by vocal pitch–taste correspondences. We used buying intention as the dependent variable in a further ANOVA, in which pitch (higher, lower) functioned as the between-participants factor and perceived taste (sweet, sour, bitter) as the within-participant factor.

The main effect of the vocal pitch of the voiceover advertisements was significant $[F(1, 98) = 11.778, p = 0.001, \eta_p^2 = 0.107]$. However, the main effect of taste was also significant $[F(1.78, 174.60) = 13.252, p < 0.001, \eta_p^2 = 0.119]$, indicating that participants had differential food preferences depending on the food taste. Post-hoc analyses indicated that the participants exhibited greater buying intention toward sweet food ($M = 3.76 \pm 1.42$) than

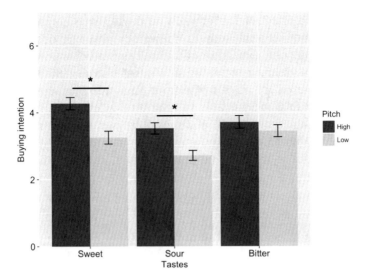

Figure 2. Results of Study 2. Effects of simplified voiceover advertisements with varying pitch on buying intention toward foods. Advertisements in a higher-pitched (vs lower-pitched) voice increased buying intention toward both sweet and sour foods. By contrast, advertisements in a higher-pitched (vs lower-pitched) voice did not influence buying intention toward bitter foods. The buying intentions were rated on a seven-point scale, with 1 = not at all and 7 = very much. Error bars show standard error. Asterisks indicate statistical significance (*, $p < 0.05$).

toward sour food ($M = 3.12 \pm 1.19$: $t_{98} = 4.777$, adjusted $p < 0.001$), and greater buying intention toward bitter food ($M = 3.58 \pm 1.31$) than sour food ($t_{98} = 4.463$, adjusted $p < 0.001$). Sweet and bitter food did not differ in terms of their buying intention ratings ($t_{98} = 1.276$, $p = 0.205$, adjusted $p = 0.205$).

Notably, there was a significant interaction between vocal pitch and taste [$F(1.78, 174.60) = 4.566$, $p = 0.015$, $\eta_p^2 = 0.045$]. A planned comparison revealed that the higher-pitched (vs lower-pitched) voiceover message increased buying intention toward sweet food [higher: $M = 4.27 \pm 1.27$ vs lower: $M = 3.25 \pm 1.38$, $F(1, 98) = 14.715$, $p < 0.001$, $\eta_p^2 = 0.131$] and sour food [higher: $M = 3.52 \pm 1.19$ vs lower: $M = 2.71 \pm 1.07$, $F(1, 98) = 12.852$, $p < 0.001$, $\eta_p^2 = 0.116$]. By contrast, the higher-pitched (vs lower-pitched) voiceover advertisement did not influence buying intention toward bitter food [higher: $M = 3.71 \pm 1.32$ vs lower: $M = 3.44 \pm 1.31$, $F(1, 98) = 1.026$, $p = 0.314$, $\eta_p^2 = 0.010$]. The results are presented in Fig. 2.

3.3. Discussion

Using a simplified message with a consistent degree of change in pitch, we found that higher- (vs lower-)pitched voiceover advertisements increased buying intention toward sour and sweet food. We replicated the results of Study 1 with respect to the effects of higher pitch on buying intention toward sour food. We found an effect of higher pitch on buying intention toward sweet food, which was not seen in Study 1. This effect may only occur when the pitch is much higher. One caveat regarding the observed change in sweet food preference is the difference in voiceover content between Studies 1 and 2. In Study 2, we altered the degree of pitch change and message contents simultaneously, and thus were unable to determine whether the level of pitch or simplified message induced the observed shift in preference toward sweet food. To solve this problem, Study 3 manipulated the message contents (simplified vs taste-arousing) and pitch (higher vs lower) simultaneously.

4. Study 3

4.1. Method

4.1.1. Design
The aim of Study 3 was to (1) replicate the effect of higher pitch on sweet food preference seen in Study 2, and (2) clarify whether the effects of higher pitch on sweet food preference were dependent on message type (taste-evoking, as used in Study 1, vs simplified, as used in Study 2). The study had a 2 (pitch: higher, lower) × 2 (message type: simplified, taste-evoking) × 3 (perceived taste: sweet, sour, bitter) mixed design, wherein pitch and message type represented the between-participants factors and taste represented the within-participant factor. The main outcome was buying intention with regard to food products.

4.1.2. Participants
In total, 202 healthy participants were recruited through Lancers (https://www.lancers.jp/) and completed the survey hosted on Qualtrics (https://www.qualtrics.com/jp/). As a study prerequisite, we asked participants to confirm that they had an audio-capable device and headphones. We excluded the data of two participants who did not wear a headset or listen to voiceover advertising. The total sample size was 200 (73 females; $M_{age} = 40.96$ years, SD = 9.74).

4.1.3. Task Procedure
We used the voiceover scripts from Studies 1 and 2, with the pitch adjusted up or down by 25%. The voiceover script in Study 1 was a taste-evoking message (e.g., "Enjoy a creamy texture and sweet vanilla flavor"). The voiceover script in Study 2 was a simplified message (e.g., "Would you like a cream puff?").

As in Study 2, we first asked participants to put on their headphones and confirm that they could hear the sound file playing. Then, participants were randomly allocated to four groups: higher pitch/simplified message ($n = 53$, 15 females), higher pitch/taste-evoking message ($n = 45$, 20 females), lower pitch/simplified message ($n = 53$, 16 females), and lower pitch/taste-evoking message ($n = 49$, 22 females).

After listening to each voiceover message, participants were asked to rate their buying intention and perception of sweet/sour/bitter taste (as in Studies 1 and 2). The participants listened to the sample voiceover script ("Would you like this product?"), and were asked to rate the degree of to which the voice matched sweet/sour/bitter taste ("How much do you think this voice is associated with sweetness/sourness/bitterness?") on a 7-point Likert scale. They also rated the valence, arousal, femininity/masculinity, pitch, and volume of the voiceover advertisement (as in Study 2).

4.2. Results

4.2.1. Pitch Perception

We tested whether the participants perceived the vocal pitch in the advertising messages differently depending on whether they listened to them under higher- or lower-pitched conditions. The participants in the higher pitch condition perceived the pitch as higher than did those in the lower pitch condition ($M_{higher} = 4.56 \pm 1.13$ vs $M_{lower} = 2.25 \pm 0.97$; $t_{198} = 15.569$, $p < 0.001$, Cohen's $d = 2.202$). The participants in the higher pitch condition perceived the masculinity of the voiceover as lower (i.e., femininity as higher) than did those in the lower pitch condition ($M_{higher} = 2.36 \pm 1.06$ vs $M_{lower} = 4.29 \pm 1.56$; $t_{198} = -10.217$, $p < 0.001$, Cohen's $d = -1.445$). Participants in the higher pitch condition perceived arousal as higher than did those in the lower pitch condition ($M_{higher} = 3.24 \pm 1.35$ vs $M_{lower} = 2.80 \pm 1.26$; $t_{198} = 2.332$, $p = 0.021$, Cohen's $d = 0.330$). Perceptions of volume ($M_{higher} = 4.17 \pm 0.63$ vs $M_{lower} = 4.31 \pm 0.56$; $t_{198} = -1.668$, $p = 0.097$) and valence (preference; $M_{higher} = 4.28 \pm 1.11$ vs $M_{lower} = 4.17 \pm 1.13$; $t_{198} = 0.688$, $p = 0.492$) did not differ between the higher and lower pitch conditions ($M_{higher} = 4.17 \pm 0.63$ vs $M_{lower} = 4.31 \pm 0.56$; $t_{198} = -1.668$, $p = 0.097$). Sweet matching ($M_{higher} = 3.83 \pm 1.42$ vs $M_{lower} = 3.85 \pm 1.42$; $t_{98} = -0.045$, $p = 0.964$), sour matching ($M_{higher} = 2.96 \pm 1.34$ vs $M_{lower} = 2.62 \pm 1.19$; $t_{98} = 1.357$, $p = 0.178$), and bitter matching ($M_{higher} = 2.58 \pm 1.43$ vs $M_{lower} = 2.42 \pm 1.24$; $t_{98} = 0.600$, $p = 0.550$) did not differ between the higher and lower pitch conditions.

4.2.2. Perception of Sweetness

We tested whether the participants perceived the sweet food as sweeter than the other foods. We used sweetness as the dependent variable in an ANOVA, in which pitch (higher, lower) and message content (simplified, taste-evoking) functioned as the between-participants factors and taste (sweet, sour, bitter) as the within-participant factor.

There was no main effect of the vocal pitch of the voiceover advertisements [$F(1, 196) = 2.567$, $p = 0.111$, $\eta_p^2 = 0.013$] or content of messages

$[F(1, 196) = 0.210, \ p = 0.647, \ \eta_p^2 = 0.001]$. However, the main effect of taste was significant $[F(1.83, 358.33) = 284.319, \ p < 0.001, \ \eta_p^2 = 0.592]$, indicating that participants perceived foods differently depending on the taste. Post-hoc analyses showed that the participants' perception of sweetness was greater for sweet food ($M = 5.07 \pm 1.37$) than for sour ($M = 2.88 \pm 1.33$: $t_{196} = 19.511$, adjusted $p < 0.001$) or bitter foods ($M = 2.58 \pm 1.32$: $t_{196} = 18.786$, adjusted $p < 0.001$). Participants also exhibited a heightened perception of sweetness for sour food than for bitter food ($t_{196} = 3.144$, adjusted $p = 0.002$). There was no significant two-way interaction between vocal pitch and taste $[F(1.83, 358.33) = 1.040, \ p = 0.350, \ \eta_p^2 = 0.005]$, and no three-way interaction of vocal pitch, content of messages, and taste $[F(1.83, 358.33) = 0.328, \ p = 0.701, \ \eta_p^2 = 0.002]$.

4.2.3. Perception of Sourness

We tested whether participants perceived the sour food as sourer than the other foods. We used sourness as the dependent variable in an ANOVA, in which pitch (higher, lower) and message content (simplified, taste-evoking) functioned as the between-participants factors, and taste (sweet, sour, bitter) as the within-participant factor.

There was no main effect of the vocal pitch of the voiceover advertisements $[F(1, 196) = 0.621, \ p = 0.432, \ \eta_p^2 = 0.003]$ or content of messages $[F(1, 196) = 1.169, \ p = 0.281, \ \eta_p^2 = 0.006]$. However, the main effect of taste was significant $[F(1.87, 366.56) = 350.039, \ p < 0.001, \ \eta^2{}_p = 0.641]$, indicating that participants perceived foods differently depending on the taste. Post-hoc analyses showed that the participants' perception of sourness was greater for sour food ($M = 5.03 \pm 1.36$) than for sweet ($M = 2.17 \pm 1.33$: $t_{196} = 22.468$, adjusted $p < 0.001$) and bitter foods ($M = 2.77 \pm 1.34$: $t_{196} = 19.887$, adjusted $p < 0.001$). Participants also exhibited a greater perception of sourness for bitter food than for sweet food ($t_{196} = 6.028$, adjusted $p = 0.002$). There was no significant two-way interaction between vocal pitch and taste $[F(1.87, 366.56) = 2.030, \ p = 0.136, \ \eta_p^2 = 0.010]$, and no three-way interaction of vocal pitch, content of messages, and taste $[F(1.87, 366.56) = 0.149, \ p = 0.848, \ \eta_p^2 = 0.0008]$.

4.2.4. Perception of Bitterness

We tested whether participants perceived the bitter food as more bitter than the other foods. We used bitterness as the dependent variable in an ANOVA, in which pitch (higher, lower) and content of messages (simplified, taste-evoking) functioned as the between-participants factors, with taste (sweet, sour, bitter) as the within-participant factor.

There was no main effect of the vocal pitch of voiceover advertisements, $F(1, 196) = 2.414, \ p = 0.122, \ \eta_p^2 = 0.012$, or content of messages

$[F(1, 196) = 0.019, p = 0.891, \eta_p^2 = 0.0001]$. However, the main effect of taste was significant $[F(1.85, 362.92) = 290.291, p < 0.001, \eta_p^2 = 0.597]$, indicating that participants perceived foods differently depending on the taste. Post-hoc analyses showed that the participants' perception of bitterness was greater for bitter food ($M = 4.85 \pm 1.46$) than for sweet ($M = 2.03 \pm 1.31$: $t_{196} = 21.344$, adjusted $p < 0.001$) and sour foods ($M = 2.85 \pm 1.44$: $t_{196} = 15.914$, adjusted $p < 0.001$). Participants also exhibited a heightened perception of bitterness for sour food than for sweet food ($t_{196} = 8.122$, adjusted $p < 0.001$). There was no significant two-way interaction between vocal pitch and taste $[F(1.85, 362.92) = 1.300, p = 0.273, \eta_p^2 = 0.007]$, and no three-way interaction of vocal pitch, content of messages, and taste $[F(1.85, 362.92) = 0.295, p = 0.728, \eta_p^2 = 0.002]$.

4.2.5. Buying Intention

We carried out experiments to determine whether the participants' food preferences were affected by vocal pitch–taste correspondences. We used buying intention as the dependent variable in a further ANOVA, in which pitch (higher, lower) and content of messages (simplified, taste-evoking) functioned as the between-participants factors and perceived taste (sweet, sour, bitter) as the within-participant factor.

There was no main effect of the vocal pitch of the voiceover advertisements $[F(1, 196) = 2.391, p = 0.124, \eta_p^2 = 0.012]$. However, the main effect of message content was significant $[F(1, 196) = 14.947, p < 0.001, \eta_p^2 = 0.071]$, indicating that participants expressed greater preference for foods following taste-evoking (vs simplified) messages. The main effect of taste was significant $[F(1, 392) = 18.140, p < 0.001, \eta_p^2 = 0.085]$, indicating that participants preferred foods to a varying extent depending on the taste. Post-hoc analyses indicated that the participants exhibited greater buying intention toward sweet food ($M = 4.06 \pm 1.53$) than toward sour food ($M = 3.58 \pm 1.38$: $t_{196} = 5.957$, adjusted $p < 0.001$), and toward bitter food ($M = 3.91 \pm 1.38$) than sour food ($t_{196} = 4.133$, adjusted $p < 0.001$). Sweet and bitter foods did not differ in terms of their buying intention ratings ($t_{196} = 1.784$, adjusted $p = 0.0760$).

Notably, there was a significant interaction between vocal pitch and taste $[F(2, 372) = 8.923, p < 0.001, \eta_p^2 = 0.044]$. A planned comparison revealed that the higher-pitched (vs lower-pitched) voiceover message increased buying intention toward sweet food [high: $M = 4.36 \pm 1.51$ vs low: $M = 3.71 \pm 1.49$, $F(1, 196) = 10.574, p = 0.001, \eta_p^2 = 0.051$]. By contrast, the higher-pitched (vs lower-pitched) voiceover advertisement did not influence buying intention toward sour food [higher: $M = 3.57 \pm 1.38$ vs lower: $M = 3.54 \pm 1.40$, $F(1, 196) = 0.034, p = 0.854, \eta_p^2 = 0.0002$] or bitter food [higher: $M = 3.96 \pm 1.41$ vs lower: $M = 3.84 \pm 1.36$, $F(1, 196) = 0.257, p = 0.613, \eta_p^2 =$

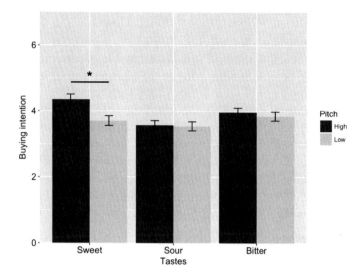

Figure 3. Results of Study 3. Effects of simplified vs taste-evoking voiceover advertisements with varying pitch on buying intention toward foods. Advertisements in a higher-pitched (vs lower-pitched) voice increased buying intention toward sweet food regardless of the message content. The bar graph was created by combining the simplified and taste-evoking conditions. Advertisements in a higher-pitched (vs lower-pitched) voice did not influence buying intention toward sour or bitter foods. The buying intentions were rated on a seven-point scale, with 1 = not at all and 7 = very much. Error bars show standard error. Asterisks indicate statistical significance (*, $p < 0.05$).

0.001]. There was no significant three-way interaction of vocal pitch, message contents and taste [$F(2, 392) = 1.346$, $p = 0.262$, $\eta_p^2 = 0.007$]. The results are presented in Fig. 3.

4.3. Discussion

The higher- (vs lower-) pitched voiceover advertisement increased buying intention toward sweet food. This effect occurred regardless of the message contents. Both taste-evoking and simplified messages enhanced buying intention toward sweet food. Thus, we replicated the results of Study 2 using different types of messages. However, we did not replicate the previous findings that higher-pitched voiceover advertisements increased buying intention toward sour food.

5. General Discussion

5.1. Summary of Findings

In recent years, various cross-modal correspondences have been reported and applied to the study of consumer behavior. The human voice is ubiquitous in marketing communication, particularly broadcast advertising (Chattopad-hyay *et al.*, 2003; Martín-Santana *et al.*, 2015). However, the issue of whether the invocation of voices and product types interactively influence consumer response to advertising has hitherto remained unconfirmed. Adopting the cross-modal correspondence phenomenon as its basis, this study investigated whether perceived taste and vocal pitch interact with one another to influence advertising effectiveness. The results show that participants were likely to ex-hibit greater buying intention toward both sweet and sour foods when they listened to higher-pitched (vs lower-pitched) voiceover advertisements. The influence of higher pitch on sweet and sour food preferences was observed in only two of the three studies: studies 1 and 2 for sour food, and studies 2 and 3 for sweet food. Taken together, these findings indicate that consumers eval-uate advertising messages for both sweet and sour food on the basis of voice pitch–taste correspondence.

These findings contribute to our understanding of how sensory factors con-tribute to preference formation. As the feeling-as-information framework sug-gests (e.g., Motoki and Sugiura, 2018; Schwarz, 2011), affective information derived from incidental experiences may carry over to other judgements. Pre-vious studies have shown that cross-modal congruency leads to increases in consumer preferences (e.g., Hagtvedt and Brasel, 2016; Motoki *et al.*, 2019b; Sunaga *et al.*, 2016; Velasco *et al.*, 2019). Our findings add to this evi-dence, showing that the cross-modal correspondence phenomenon can apply to voiceover advertisements.

This study demonstrated that a pitch manipulation may lead to greater preference for sweet food under certain circumstances. When pitch was mod-erately heightened (increased by 20% compared to the original sound), it did not influence sweet food preferences (Study 1). However, when the pitch was much higher (25% increase compared to the original sound), it enhanced sweet food preferences (Studies 2 and 3). Importantly, the ef-fects of higher pitch on buying intention toward sweet food occurred re-gardless of message content. That is, whether the message was short (only product words) or long (arousing information), a higher-pitched voice in-creased the preference for sweet food. Together, these findings suggest that a much higher vocal pitch is effective for voiceover advertisements for sweet foods.

The F0 of the higher vocal pitch used in Studies 2 and 3 is in the consid-erably high range, i.e., about 300 Hz (Appendix Tables A2 and A3). The F0 for the higher-pitched voice, leading to enhanced sweet food preference, is beyond the ordinary female vocal range (200–240 Hz; e.g., Borkowska and Pawlowski, 2011; Coleman, 1976; Lattner *et al.*, 2005). The vocal pitch is also higher than a voice associated with the traditional notion of Japanese femininity (262.7 Hz; Starr, 2015). Rather, it is similar to the frequency of a very high young female voice (310.3 Hz), as opposed to a low (184.6 Hz), medium (223.7 Hz), or high (261.9 Hz) young female voice (Borkowska and Pawlowski, 2011). The vocal pitch of the higher-pitched voice was also similar to that of a voice type commonly used for animated female characters (320 and 400 Hz; Teshigawara, 2003). The vocal pitch that leads to enhanced sweet food preference may correspond to the very high voice of a young woman or the voices of anime characters.

5.2. Possible Mechanisms

Given the previous findings (Crisinel *et al.*, 2012b; Reinoso Carvalho *et al.*, 2016, 2017; Wang and Spence, 2016; Wang *et al.*, 2017; Watson and Gunther, 2017), it is apparently possible that perceived taste may mediate the associations between vocal pitch and buying intentions. Study 1 demonstrated that higher (vs lower) vocal pitch reduced the perception of sour tastes, which increases buying intention, thereby indicating a possible role of shifts in perceived sourness with regard to increasing buying intention. However, we did not find any evidence supporting the role of perceived sourness in Studies 2 and 3. Although higher (vs lower) vocal pitch increased the buying intention in Study 2, it did not decrease the perceived sourness in Studies 2 and 3. The possible mechanisms remain unknown. Further studies will be necessary to investigate this issue.

The possible mechanisms leading to associations between vocal pitch and buying intention for sweet foods remain unknown. Although it is apparently possible that manipulation of vocal pitch increases buying intention by shifting perceived taste, the results were not as expected. Although Studies 2 and 3 demonstrated that higher (vs lower) vocal pitch increases buying intention for sweet foods, only the femininity of the voices increased consistently with higher-pitched (vs lower-pitched) voiceovers. Consequently, to ascertain the possible mechanisms, we analyzed the correlation between femininity and buying intention for sweet foods. Femininity and buying intentions for sweet foods were positively correlated in Study 3 ($r = -0.161$, $p = 0.022$), but not in Study 2 ($r = -0.101$, $p = 0.303$). Although the femininity of the voice may be a possible determining factor, the evidence is still preliminary. Further study should be carried out in order to clarify the details of the possible mechanisms.

5.3. Why Were No Effects of Vocal Pitch Observed in Relation to Bitter Foods?

Although the correspondences between sweet/sour taste and a higher-pitched voice influenced advertising effectiveness, the other possible correspondences (i.e., bitterness/lower-pitched voice) did not. The null effect observed with regard to lower-pitched voice/bitterness may be attributed to the degree of low pitch. Although the participants perceived the lower-pitched voices as lower than the higher-pitched voices, the lower-pitched voices were conveyed by women. If the lower-pitched voice had been produced by a man and the pitch had been much lower, effects of voice pitch on preference for bitter food may have been observed. Further analysis is required to clarify these issues through more precise control of the degree of vocal pitch.

5.4. Limitations

First, it is difficult to conclude whether the effects observed in this study were driven by absolute or relative pitch. It has been shown that the effects of auditory/visual senses on cognitive performance are driven by relative rather than absolute stimuli (Brunetti *et al.*, 2018; Spence, 2019). In view of this, the results of the present study may have been affected by relative pitch differences. Second, the results were somewhat inconsistent for sour food. A higher-pitched voice increased preference for sour food (Studies 1 and 2), but this effect was not replicated in Study 3. Further study is needed to confirm the replicability of the effects using larger sample sizes. Third, the effects of a higher-pitched voice (as opposed to a lower-pitched one) on sweet/sour food preference appears to be relatively small. However, broadcast advertisements are among the most cost-intensive forms of marketing communication. Billions of dollars have been invested in broadcast advertising and, thus, even the subtlest improvements in advertising effectiveness are likely to be deemed of value. Fourth, the characteristics of foods differ in terms of preference, tastiness, and healthiness. For example, the participants preferred sweet foods (cream puffs) to sour foods (lemon juice) and bitter foods (black coffee). This fact might influence the results. Further studies should be carried out to address this issue. Finally, it is not clear that the change in pitch itself influences the buying intention for sweet/sour foods. Formants and spectral balance (Simner *et al.*, 2010) as well as timbre (Crisinel and Spence, 2010) have some associations with taste, because the pitch change, formants, and spectral balance vary in a similar manner, so they may influence the results. Once again, further study should be undertaken to investigate these related issues.

5.5. Recommendations for Future Study

The investigation of whether voice tempo and taste interactively influence consumer responses to advertising should be prioritized. There are further acoustic features of voice that were beyond the scope of this study, tempo among them. Cross-modal correspondences between sound tempo and taste have been demonstrated (Bronner *et al.*, 2012; Knoeferle *et al.*, 2015). For example, slow tempo is reliably associated with sweet taste, while fast tempo reliably corresponds with sour taste (Bronner *et al.*, 2012). High pitch tends to be associated with fast tempo, and low pitch with slow tempo (Wang and Spence, 2017). Future study should investigate whether voice tempo influences preference for advertised foods.

5.6. Practical Contributions

This study's findings increase the applicability of voice pitch–taste congruency to industry. When marketers select voiceover actors or actresses to advertise sour or sweet foods, they may find it worthwhile to consider the speakers' vocal pitches. Generally, women's voices are higher in pitch than men's; thus, an advertisement featuring a female voice may be more effective in promoting a sour or sweet food product. Moreover, even within the same gender, vocal pitch varies. Buying intention toward sweet food increased when the voice pitch was considerably high. As such, when marketers have decided to recruit a female voiceover actor to advertise a sweet food product, they should narrow their search further to women with higher vocal pitches. Additionally, assuming that pitch does play a critical role in sour or sweet food preference, background music with a higher pitch may increase positive attitudes to sour or sweet food advertisements. These are the implications based on this study's findings.

5.7. Conclusion

This is the first study to verify the role of voice pitch–taste correspondence in broadcast advertising. In particular, the findings reveal that a higher-pitched voice increases preference for sour and sweet food. This work emphasizes the importance of the relationship between the voiceover actor's voice and the advertised food product with regard to consumer preference. In conclusion, these findings suggest that voice pitch and taste interact to influence advertising effectiveness and, thereby, this study has advanced the applicability of cross-modal correspondences to business.

Acknowledgements

This study was supported by JSPS KAKENHI Grant Number 17J00389 (K.M.), KAKENHI Grant Number 16H01873 from MEXT (M.S.), and a Grant-in-Aid for Scientific Research on Innovative Areas (Research in a Proposed Research Area: 17H06046) (R.N.). The authors would like to thank the editor and anonymous reviewers for their helpful comments, which significantly improved the final version of the paper.

References

Biswas, D., Lund, K. and Szocs, C. (2019). Sounds like a healthy retail atmospheric strategy: effects of ambient music and background noise on food sales, *J. Acad. Mark. Sci.* **47**, 37–55.

Borkowska, B. and Pawlowski, B. (2011). Female voice frequency in the context of dominance and attractiveness perception, *Anim. Behav.* **82**, 55–59.

Bremner, A. J., Caparos, S., Davidoff, J., de Fockert, J., Linnell, K. J. and Spence, C. (2013). "Bouba" and "Kiki" in Namibia? A remote culture make similar shape–sound matches, but different shape–taste matches to Westerners, *Cognition* **126**, 165–172.

Bronner, K., Frieler, K., Bruhn, H., Hirt, R. and Piper, D. (2012). What is the sound of citrus? Research on the correspondences between the perception of sound and flavour, in: *Proceedings of the 12th International Conference of Music Perception and Cognition (ICMPC) and the 8th Triennial Conference of the European Society for the Cognitive Sciences of Music (ESCOM) 2012*, pp. 142–148. Thessaloniki, Greece.

Brunetti, R., Indraccolo, A., Del Gatto, C., Spence, C. and Santangelo, V. (2018). Are cross-modal correspondences relative or absolute? Sequential effects on speeded classification, *Atten. Percept. Psychophys.* **80**, 527–534.

Chattopadhyay, A., Dahl, D. W., Ritchie, R. J. B. and Shahin, K. N. (2003). Hearing voices: the impact of announcer speech characteristics on consumer response to broadcast advertising, *J. Consum. Psychol.* **13**, 198–204.

Chebat, J.-C., Hedhli, K. E. L., Gélinas-Chebat, C. and Boivin, R. (2007). Voice and persuasion in a banking telemarketing context, *Percept. Mot. Skills* **104**, 419–437.

Coleman, R. O. (1976). A comparison of the contributions of two voice quality characteristics to the perception of maleness and femaleness in the voice, *J. Speech Hear. Res.* **19**, 168–180.

Crisinel, A.-S. and Spence, C. (2011). A fruity note: crossmodal associations between odors and musical notes, *Chem. Sens.* **37**, 151–158.

Crisinel, A.-S. and Spence, C. (2009). Implicit association between basic tastes and pitch, *Neurosci. Lett.* **464**, 39–42.

Crisinel, A.-S. and Spence, C. (2010). As bitter as a trombone: synesthetic correspondences in nonsynesthetes between tastes/flavors and musical notes, *Atten. Percept. Psychophys.* **72**, 1994–2002.

Crisinel, A.-S., Jones, S. and Spence, C. (2012a). 'The sweet taste of maluma': crossmodal associations between tastes and words, *Chemosens. Percept.* **5**, 266–273.

Crisinel, A.-S., Cosser, S., King, S., Jones, R., Petrie, J. and Spence, C. (2012b). A bittersweet symphony: systematically modulating the taste of food by changing the sonic properties of the soundtrack playing in the background, *Food Qual. Pref.* **24**, 201–204.

Deroy, O., Crisinel, A.-S. and Spence, C. (2013). Crossmodal correspondences between odors and contingent features: odors, musical notes, and geometrical shapes, *Psychon. Bull. Rev.* **20**, 878–896.

Emarketers (2018a). US Digital Video Ad Spending, 2017–2021 (billions, % change and % of total digital ad spending). https://www.emarketer.com/Chart/US-Digital-Video-Ad-Spending-2017-2021-billions-change-of-total-digital-ad-spending/209000. Retrieved 3 October 2018.

Emarketers (2018b). US TV Ad Spending to Fall in 2018. https://www.emarketer.com/content/us-tv-ad-spending-to-fall-in-2018. Retrieved 3 October 2018.

Hagtvedt, H. and Brasel, S. A. (2016). Cross-modal communication: sound frequency influences consumer responses to color lightness, *J. Mark. Res.* **53**, 551–562.

Ho, H. N., Van Doorn, G. H., Kawabe, T., Watanabe, J. and Spence, C. (2014). Colour-temperature correspondences: when reactions to thermal stimuli are influenced by colour, *PloS One* **9**, e91854. DOI:10.1371/journal.pone.0091854.

Knoeferle, K., Li, J., Maggioni, E. and Spence, C. (2017). What drives sound symbolism? Different acoustic cues underlie sound–size and sound–shape mappings, *Sci. Rep.* **7**, 5562. DOI:10.1038/s41598-017-05965-y.

Knoeferle, K. M., Woods, A., Käppler, F. and Spence, C. (2015). That sounds sweet: using cross-modal correspondences to communicate gustatory attributes, *Psychol. Mark.* **32**, 107–120.

Knöferle, K. and Spence, C. (2012). Crossmodal correspondences between sounds and tastes, *Psychon. Bull. Rev.* **19**, 992–1006.

Lattner, S., Meyer, M. E. and Friederici, A. D. (2005). Voice perception: sex, pitch, and the right hemisphere, *Hum. Brain Mapp.* **24**, 11–20.

Ligges, U., Preusser, A., Thieler, A., Mielke, J. and Weihs, C. (2018). 'tuneR: analysis of Music and Speech'. R package version 1.3.3. ftp://ctan.uib.no/pub/cran/web/packages/tuneR/tuneR.pdf. Retrieved 01 January 2001.

Martín-Santana, J. D., Muela-Molina, C., Reinares-Lara, E. and Rodríguez-Guerra, M. (2015). Effectiveness of radio spokesperson's gender, vocal pitch and accent and the use of music in radio advertising, *Bus. Res. Q.* **18**, 143–160.

Motoki, K. and Sugiura, M. (2018). Disgust, sadness, and appraisal: disgusted consumers dislike food more than sad ones, *Front. Psychol.* **9**, 76. DOI:10.3389/fpsyg.2018.00076.

Motoki, K., Saito, T., Nouchi, R., Kawashima, R. and Sugiura, M. (2018). The paradox of warmth: ambient warm temperature decreases preference for savory foods, *Food Qual. Pref.* **69**, 1–9.

Motoki, K., Saito, T., Nouchi, R., Kawashima, R. and Sugiura, M. (2019a). Round faces are associated with sweet foods: the role of crossmodal correspondence in social perception, *Foods* **8**, 103. DOI:10.3390/foods8030103.

Motoki, K., Saito, T., Nouchi, R., Kawashima, R. and Sugiura, M. (2019b). Light colors and comfortable warmth: crossmodal correspondences between thermal sensations and color lightness influence consumer behavior, *Food Qual. Pref.* **72**, 45–55.

Ngo, M. and Spence, C. (2011). Assessing the shapes and speech sounds that people associate with different kinds of chocolate, *J. Sens. Stud.* **26**, 421–428.

Ngo, M. K., Misra, R. and Spence, C. (2011). Assessing the shapes and speech sounds that people associate with chocolate samples varying in cocoa content, *Food Qual. Pref.* **22**, 567–572.

North, A. C., Hargreaves, D. J. and McKendrick, J. (1999). The influence of in-store music on wine selections, *J. Appl. Psychol.* **84**, 271–276.

Oakes, S. (2007). Evaluating empirical research into music in advertising: a congruity perspective, *J. Advert. Res.* **47**, 38–50.

Puts, D. A., Gaulin, S. J. C. and Verdolini, K. (2006). Dominance and the evolution of sexual dimorphism in human voice pitch, *Evol. Hum. Behav.* **27**, 283–296.

Ramachandran, V. S. and Hubbard, E. M. (2001). Synaesthesia — a window into perception, thought and language, *J. Consc. Stud.* **8**, 3–34.

Reinoso Carvalho, F., Wang, Q. (J.), Van Ee, R. and Spence, C. (2016). The influence of sound-scapes on the perception and evaluation of beers, *Food Qual. Pref.* **52**, 32–41.

Reinoso Carvalho, F., Wang, Q. (J.), van Ee, R., Persoone, D. and Spence, C. (2017). "Smooth operator": music modulates the perceived creaminess, sweetness, and bitterness of chocolate, *Appetite* **108**, 383–390.

Schwarz, N. (2011). Feelings-as-information theory, in: *The Handbook of Theories of Social Psychology*, P. A. M. Van Lange, A. Kruglanski and E. T. Higgins (Eds), pp. 289–308. Sage, Thousand Oaks, CA, USA.

Shue, Y.-L., Keating, P., Vicenik, C. and Yu, K. (2011). VoiceSauce: a program for voice analysis, in: *Proceedings of the 17th ICPhS, Hong Kong*, pp. 1846–1849.

Simmons, J. P., Nelson, L. D. and Simonsohn, U. (2011). False-positive psychology: undisclosed flexibility in data collection and analysis allows presenting anything as significant, *Psychol. Sci.* **22**, 1359–1366.

Simmons, J. P., Nelson, L. D. and Simonsohn, U. (2018). False-positive citations, *Perspect. Psychol. Sci.* **13**, 255–259.

Simner, J., Cuskley, C. and Kirby, S. (2010). What sound does that taste? Cross-modal mappings across gustation and audition, *Perception* **39**, 553–569.

Spence, C. (2011). Crossmodal correspondences: a tutorial review, *Atten. Percept. Psychophys.* **73**, 971–995.

Spence, C. (2012). Managing sensory expectations concerning products and brands: capitalizing on the potential of sound and shape symbolism, *J. Consum. Psychol.* **22**, 37–54.

Spence, C. (2019). On the relative nature of (pitch-based) crossmodal correspondences, *Multisens. Res.* **32**, 235–265. DOI:10.1163/22134808-20191407.

Spence, C. and Gallace, A. (2011). Tasting shapes and words, *Food Qual. Pref.* **22**, 290–295.

Spence, C., Puccinelli, N. M., Grewal, D. and Roggeveen, A. L. (2014). Store atmospherics: a multisensory perspective, *Psychol. Mark.* **31**, 472–488.

Starr, R. L. (2015). Sweet voice: the role of voice quality in a Japanese feminine style, *Lang. Soc.* **44**, 1–34.

Stel, M., van Dijk, E., Smith, P. K., van Dijk, W. W. and Djalal, F. M. (2012). Lowering the pitch of your voice makes you feel more powerful and think more abstractly, *Soc. Psychol. Personal. Sci.* **3**, 497–502.

Sunaga, T., Park, J. and Spence, C. (2016). Effects of lightness-location congruency on consumers' purchase decision-making, *Psychol. Mark.* **33**, 934–950.

Teshigawara, M. (2003). *Voices in Japanese Animation: a Phonetic Study of Vocal Stereotypes of Heroes and Villains in Japanese Culture*. PhD thesis, University of Victoria, Victoria, BC, Canada.

Velasco, C., Salgado-Montejo, A., Marmolejo-Ramos, F. and Spence, C. (2014). Predictive packaging design: tasting shapes, typefaces, names, and sounds, *Food Qual. Pref.* **34**, 88–95.

Velasco, C., Woods, A. T., Petit, O., Cheok, A. D. and Spence, C. (2016). Crossmodal correspondences between taste and shape, and their implications for product packaging: a review, *Food Qual. Pref.* **52**, 17–26.

Velasco, C., Adams, C., Petit, O. and Spence, C. (2019). On the localization of tastes and tasty products in 2D space, *Food Qual. Pref.* **71**, 438–446.

Wang, Q. (J.) and Spence, C. (2016). 'Striking a sour note': assessing the influence of consonant and dissonant music on taste perception, *Multisens. Res.* **29**, 195–208.

Wang, Q. J. and Spence, C. (2017). The role of pitch and tempo in sound-temperature cross-modal correspondences, *Multisens. Res.* **30**, 307–320.

Wang, Q. (J.), Woods, A. and Spence, C. (2015). "What's your taste in music?" A comparison of the effectiveness of various soundscapes in evoking specific tastes, *i-Perception* **6**, 2041669515622001. DOI:10.1177/2041669515622001.

Wang, Q. (J.), Keller, S. and Spence, C. (2017). Sounds spicy: enhancing the evaluation of piquancy by means of a customised crossmodally congruent soundtrack, *Food Qual. Pref.* **58**, 1–9.

Watson, Q. J. and Gunther, K. L. (2017). Trombones elicit bitter more strongly than do clarinets: a partial replication of three studies of Crisinel and Spence, *Multisens. Res.* **30**, 321–335.

Appendix

Table A1.
Voice parameters of taste-evoking messages for Study 1

		Bitter		Sour		Sweet	
		High	Low	High	Low	High	Low
Formants	F0	292	223	265	213	279	194
	F1	501	512	533	493	531	509
	F2	1574	1784	1565	1730	1657	1773
Spectral balance		389	293	423	317	398	299

The pitch of the higher pitch condition was increased by 20%, and that of the lower pitch condition was decreased by 30%, compared to the original voice. The messages of voiceover advertisements were taste-evoking (e.g., "Enjoy a creamy texture and sweet vanilla flavor"). Values are given in Hz.

Table A2.
Voice parameters of simplified messages for Studies 2 and 3

		Bitter		Sour		Sweet	
		High	Low	High	Low	High	Low
Formants	F0	284	170	274	162	299	174
	F1	536	515	557	489	530	486
	F2	1573	1629	1612	1609	1463	1676
Spectral balance		501	304	522	297	437	261

Pitch was changed by 25% in both higher and lower pitch conditions. The messages of voiceover advertisements were simplified (e.g., "Would you like a cream puff?"). Values are given in Hz.

Table A3.
Voice parameters of taste-evoking messages for Study 3

		Bitter		Sour		Sweet	
		High	Low	High	Low	High	Low
Formants	F0	299	177	323	179	312	192
	F1	511	474	512	464	548	462
	F2	1523	1667	1530	1655	1578	1670
Spectral balance		438	262	477	290	446	269

Pitch was changed by 25% in both the higher and lower pitch conditions. The messages of voiceover advertisements were taste-evoking (e.g., "Enjoy a creamy texture and sweet vanilla flavor"). Values are given in Hz.

Taste the Bass: Low Frequencies Increase the Perception of Body and Aromatic Intensity in Red Wine

Jo Burzynska [1,*], **Qian Janice Wang** [2,3], **Charles Spence** [3] and
Susan Elaine Putnam Bastian [4]

[1] UNSW Art and Design, Sydney, Australia
[2] Department of Food Science, Aarhus University, Årslev, Denmark
[3] Crossmodal Research Laboratory, Oxford University, UK
[4] The University of Adelaide, School of Agriculture, Food and Wine, Glen Osmond, Australia

Abstract

Associations between heaviness and bass/low-pitched sounds reverberate throughout music, philosophy, literature, and language. Given that recent research into the field of cross-modal correspondences has revealed a number of robust relationships between sound and flavour, this exploratory study was designed to investigate the effects of lower frequency sound (10 Hz to 200 Hz) on the perception of the mouthfeel character of palate weight/body. This is supported by an overview of relevant cross-modal studies and cultural production. Wines were the tastants — a New Zealand Pinot Noir and a Spanish Garnacha — which were tasted in silence and with a 100 Hz (bass) and a higher 1000 Hz sine wave tone. Aromatic intensity was included as an additional character given suggestions that pitch may influence the perception of aromas, which might presumably affect the perception of wine body. Intensity of acidity and liking were also evaluated. The results revealed that the Pinot Noir wine was rated as significantly fuller-bodied when tasted with a bass frequency than in silence or with a higher frequency sound. The low frequency stimulus also resulted in the Garnacha wine being rated as significantly more aromatically intense than when tasted in the presence of the higher frequency auditory stimulus. Acidity was rated considerably higher with the higher frequency in both wines by those with high wine familiarity and the Pinot Noir significantly better liked than the Garnacha. Possible reasons as to why the tones used in this study affected perception of the two wines differently are discussed. Practical application of the findings are also proposed.

Keywords

Low frequency, wine body, bass, sonic seasoning, cross-modal correspondences, sound

---

[*] To whom correspondence should be addressed. E-mail: jo@joburzynska.com

1. Introduction

From Pythagoras' alleged observation that heavier hammers make lower sounds (see Note 1) to the call by promoters of the DMZ dubstep night to "come meditate on bass weight" (Note 2); low-pitched sounds in the bass register (10 Hz to 200 Hz, see Leventhall *et al.*, 2003, p. 7) have regularly been associated with ideas of weight, heaviness, and thickness. In terms of research on cross-modal correspondences (see Parise, 2016; Spence, 2011, for reviews), the results of a study by Walker *et al.* (2017) demonstrated that heavier objects tend to cross-activate features of lower-pitched sounds; while pitch has been identified as a sonic parameter that can be consistently conceptually, or perceptually, mapped on to tastes/flavours (e.g., Bronner *et al.*, 2012; Crisinel and Spence, 2010a; Knoeferle *et al.*, 2015; Kontukoski *et al.*, 2015; Mesz *et al.*, 2011, 2012; Reinoso-Carvalho *et al.*, 2016a; Wang *et al.*, 2016) and odours (Belkin *et al.*, 1997; Crisinel and Spence, 2011; Deroy *et al.*, 2013). Taken together, such findings suggest that pitch might have an effect on somatosensory perception more widely, and potentially extend to the feeling of the weight of a drink (or perhaps even food) in the mouth, known as oral-somatosensory perception (see Spence and Piqueras-Fiszman, 2016, for a review). Certainly, the vestibular system — the sensory system responsible for balance and spatial orientation — has been shown to be highly sensitive to low sound frequencies, even at relatively low volumes (McAngus Todd *et al.*, 2008). In the exploratory experiment reported here, we hypothesized that a bass sound frequency (low pitch) might be able to alter the perception of the weight (body) of a wine. We aimed to investigate this through participants rating a wine's weight, along with a selection of other wine attributes, under different sound conditions. We additionally investigated aroma intensity, given that earlier research has indicated that pitch appears to influence the perception of aromas, and given suggestions that body can be influenced by wine aromas (Jackson, 2009, p. 154; see also Spence, 2019b). Acidity (sourness) is also included as a reference for comparison, given that this property has already been robustly associated with higher pitches (Bronner *et al.*, 2012; Crisinel and Spence, 2010a, 2012; Kontukoski *et al.*, 2015; Mesz *et al.*, 2011, 2012; Wang and Spence, 2015), as well as liking in order to examine possible hedonic influence.

1.1. Audio-Gravitational Arts

Before modern science started to tease out the details of the complex interactions between the senses, philosophers, artists, and musicians were already making connections between low pitch and weight. As reported by Zibkowski (2002, pp. 7–8), the ancient Greek philosopher Pythagoras was said to have developed his theory of musical tuning through watching blacksmiths at work and noting that the lowest sounds resonated from their heavier hammers. Gravity's intersensory associations with musical pitch are further traced in the synesthetic theories of Galayev (2003). He notes their manifestation in terms such as 'baritone' — the male voice lying between tenor and bass — whose etymological root lies in the Greek for 'heavy sound'. That low pitch should be regarded as 'heavy' and high pitch 'light' is "fixed in many languages, as well as conventional musicological terminology", and is linked to physical laws where the tendency is for low-pitched sounds to be generated from large and "therefore heavy" objects and high ones from small, light ones (Galayev, 2003, pp. 130–131). While the dominant Western notion of pitch is characterized by spatial verticality (Deroy *et al.*, 2018), inspiring Western art music's rising notation from low to high — which has been linked to natural auditory scene statistics (Parise *et al.*, 2014) — bass is nevertheless widely associated with heaviness. This is illustrated by the English Oxford Living Dictionary (2018) definition of heaviness in relation to rock music, as *"the quality of having a strong bass component"*. In other cultures, pitch is more closely encoded to weight or thickness in language. For example, in the Farsi, Turkish, and Zapotec languages, low pitch is understood metaphorically as 'thick', while high pitch is 'thin' (Shayan *et al.*, 2011). Both height–pitch and thickness–pitch associations would appear to be pre-linguistic and possibly also universal (e.g., see Dolscheid *et al.*, 2014). The ancient Greeks talked of 'sharpness' and 'heaviness' and in Bali and Java (islands in the Indonesian archipelago), pitches are described as 'small' and 'large', respectively (see Zbikowski, 1998).

Ideas of heaviness resonate through the language used by, and applied to, bass-driven musical genres, such as Reggae, Drum 'n' Bass, and Dubstep. In these, the dominant bass element is regularly described as being 'heavy', with the term applied to basslines, track titles (such as dub reggae pioneer King Tubby's "A Heavy Dub"), and band names (such as the drum 'n' bass act Delta Heavy). It is possible that this language could be linked to the physical feeling of bass. In his investigation of the creative use of bass, Jasen (2017) notes the particular capacity of bass frequencies to provoke vestibular responses, which include bodily sensations such as 'heaviness' (Jasen, 2017, p. 110). Jasen evokes the visceral nature of a genre such as dubstep, in which bass is a "vibrational force" that fills, or even overfills bodies on the dancefloor (pp. 180–181). In Henriques' (2003, 2011) sense-led exploration of the "bass culture" of the Reggae sound system, the reggae dancehall session is described as not only engaging audition, but as a "multi-sensory" (p. 20) and corporeal (p. 101) experience, with a "blood pulse" and "powerful low frequencies [that] resonate with embodied movement" (pp. 13–14).

In literature, low pitch and weight have often been yoked. In the case of poetry, for instance, a study by Macdermott (1940) analyzed almost 200 poems and discovered cross-modal sound symbolism between low-pitched vowels and sensory qualities of heaviness. Meanwhile, in Huysman's (1884/1959) novel *Against Nature/À Rebours*, within the orchestra of the palate emanating from the drinks cabinet that the protagonist calls his "mouth organ", the double-bass is described as "*full-bodied*, solid and dark as the old bitters" (p. 59, our italics).

Through the multisensory arts practice of this paper's first author, it has also been noted that bass tones appeared to boost the impression of body when creating 'oenosonic' (Note 3) compositions — such as *Oenosthesia* (Burzynska, 2018a). These observations were reinforced by responses from participants attending her wine sound matching workshops, such as the "Pinosthesia" presentation made at the Pinot Noir 2017 conference in Wellington, New Zealand in January 2017. At this event, a single Pinot Noir (Note 4) was tasted by those attending the conference with a varied selection of music. This included a composition that Burzynska had created specifically for the wine and a bass-heavy track from the dubstep musical genre (Datsik, 2009). The latter musical work elicited a number of observations similar to her own amongst the participants, in that it appeared to increase the perception of wine body. This finding was conveyed both through informal feedback and coverage of the tasting (Note 5) (D'Amato, 2017; see also Moran, 2017).

1.2. *Cross-Modal Pitch–Weight Connections*

Low-pitch and weight associations have begun to be investigated in cross-modal (or cross-sensory) science. In a recent study, subtitled "Heaviness is dark and low-pitched", Walker *et al.* (2017) observed that participants lifting unseen objects that were identical in size but varied in terms of their weight expected the heavier object to make lower pitched sounds. This reinforced the association between pitch and weight, suggesting that it exists both as a metaphor, and a perceptual, non-visual and non-linguistic phenomenon, which could be modality-independent (see Walker *et al.*, 2012, for a discussion).

In light of recent research into cross-modal correspondences between sound and taste/flavour (flavour in this paper including mouthfeel characteristics, see Spence and Piqueras-Fiszman, 2016, for a review), the potential for finding correspondences between pitch and body would appear promising. It has now been widely demonstrated that people map taste/flavour characters with both musical and non-musical sounds (e.g., see Knöferle and Spence, 2012). Musical sounds include notes, instruments (Crisinel and Spence, 2009, 2010a, 2012) and music and soundscape compositions (Bronner *et al.*, 2012; Knoeferle *et al.*, 2015; Kontukosi *et al.*, 2015; Mesz *et al.*, 2011; Reinoso-Carvalho *et al.*, 2016b; Spence *et al.*, 2013; Wang and Spence, 2015, 2018; Wang *et al.*, 2015), while non-musical sounds include pure tones (Holt-Hansen, 1968, 1976; Reinoso-Carvalho *et al.*, 2016a; Wang *et al.*, 2016), eating sounds

(Zampini and Spence, 2004; Spence, 2015, for a review) and noise (Spence, 2014; Yan and Dando, 2015). Furthermore, these matches can be more than merely cross-modal associations, with sounds sometimes being shown to modulate the perception of flavours as well (Crisinel *et al.*, 2012). Numerous studies have now robustly demonstrated that high pitches are mapped on to sour tastes (Bronner *et al.*, 2012; Crisinel and Spence, 2010b; Kontukoski *et al.*, 2015; Mesz *et al.*, 2011, 2012; Simner *et al.*, 2010; Wang and Spence, 2015). A number of cross-modal correspondences between sounds (pitch) and odours have also been identified (Belkin *et al.*, 1997; Crisinel and Spence, 2011).

In recent years, a number of studies have identified that cross-modal correspondences exist between sound and wine (for reviews of the area, see Spence, 2011; Spence and Wang 2015a, b, c, and, for a more general multisensory overview, Spence, 2019a). A number of researchers have demonstrated a high level of agreement in the congruency between pieces of music and wines in forced choice matching exercises (Spence and Wang, 2015b, c; Spence *et al.*, 2013; Wang and Spence, 2015), and that emotional associations might be mediating these sound-flavour correspondences (Wang and Spence, 2015). Only a few of these studies have included body. For instance, Wang and Spence (2018) found that a staccato soundtrack increased the perception of body amongst wine professionals. Body is an oral somatosensory wine character that is, as yet, still not fully understood (Laguna *et al.*, 2017; Niimi *et al.*, 2017), but likely resulting from the perception of a combination of alcohol, polysaccharides, sugar, and tannins in a wine (see Gawel, 2006; Gawel *et al.*, 2018). As defined by Grainger (2009, p. 136): "*Body, sometimes referred to as weight or mouth feel, is more of a tactile than a taste sensation. It is a loose term to describe the lightness or fullness of the wine in the mouth.*" No studies have specifically investigated body in relation to pitch. The existence of a consensual correspondence between pitch and specific wine aromas was established by Crisinel and Spence (2011). This study presented aromas commonly found in wine, sourced from single presentations or blends of typical molecules in the educational *Le Nez du Vin* wine aroma kit. It found participants matched much lower pitched sounds with aromas of smoke, musk, dark chocolate, and cut hay, and higher pitches with fruitier aromas. There is currently no cross-modal research that has focused on specific sound–aroma matches using actual wine.

Formal studies of sound–flavour correspondences have also yet to focus on sound stimuli in the bass frequency register. Those studies that have used lower pitches have indicated that bitter tastes are regularly matched with lower-pitched sounds (Crisinel and Spence, 2010a; Crisinel *et al.*, 2012; Knöferle and Spence, 2012). Meanwhile, a study by Wang *et al.* (2015) that used a number of different compositions associated with different basic tastes, found the best-matched 'bitter' composition had the lowest pitch of all the soundscapes that were tested.

According to other studies using alcoholic beverages, beers that had bitter profiles matched with significantly lower pitch ranges when participants were given free range to 'tune' a selection of beers across a spectrum of sine tones (Reinoso Carvalho *et al.*, 2016a). Furthermore, a putatively bitter (lower pitched) soundtrack was also found to result in people rating a beer as tasting more bitter, and alcoholic (Reinoso Carvalho *et al.*, 2016b). It is worth noting that alcohol is a component that plays a key role in the perception of body in alcoholic drinks. This study follows the earlier experiments of Holt-Hansen (1968), which demonstrated for the first time that the matches of two different beers consistently fell within two different narrow frequency bands, replicated by Rudmin and Cappelli (1983). Crucially, no studies with wine and sound reported to date have observed the effect of single frequency sine waves on the characteristics of wine, and have instead used more complex soundtracks/musical pieces with variables that make it difficult to isolate the specific elements that may have been responsible for the perceptual taste/flavour changes. In this experiment, these variables have been reduced through the sole usage of pure tones.

2. Materials and Methods

2.1. Participants — Main Study

A total of 50 participants took part in the main study: 25 at Oxford University's Crossmodal Research Laboratory in Oxford, UK, in October 2017, and another 25 at the University of New South Wales' faculty of Art and Design in Sydney, Australia, in March 2018. This sample size is supported by data gathered by Gacula and Rutenbeck (2006), confirming that 40–100 participants appear to be the ideal sample size for analytical consumer sensory tests. Both groups of participants took part in the same study using identical wines. The study was covered by ethics approval from UNSW's Human Research Ethics Committee (HC17727) and the University of Oxford's Central University Research Ethics Committee (MSD-IDREC-C1-2014-205). All of the participants were aged over 18 years of age, and gave their informed consent prior to taking part in the study. The participants also confirmed that they met the study's criteria: that they did not suffer or have suffered from any taste, olfactory or auditory dysfunction; were not aware of possessing a sensitivity to low frequencies (Note 6); were not currently suffering from a cold/flu, or other temporary respiratory problems, and had no current or past alcohol dependency issues.

Of the 50 participants, 18 were male and 32 female. They were aged between 22 and 73 years (mean age = 41.18 years, SD = 13.23). In terms of self-reported wine expertise; 6 had no expertise, 13 were beginners, 16 intermediate, and 15 were advanced, which when combined was 19 novice/beginners and 31 with some experience. In terms of tasting frequency, one drank wine up to six times a year, 10 drank wine monthly, 19 weekly, and 20 drank wine most days. When combined, 30 were relatively infrequent wine drinkers and 20 were frequent.

2.2. Stimuli

2.2.1. Auditory Stimuli
The wine was tasted in silence and with a low and a higher pure tone presented in a randomized order. These tones were sine waves with a bass frequency of 100 Hz (approximately the musical note G_2), and, as a contrast, a higher 1000 Hz frequency sine wave (approximately the musical note B_5). These were presented at a comfortable headphone level similar to that of everyday speech, with scaled sound pressure levels derived from the IS0 226:2003 Equal Loudness Contours (ISO, 2003): the 100 Hz at 68 dB, and the 1000 Hz at 60 dB. The auditory stimuli were downloaded from the wavTones.com Professional Online Audio Frequency Signal Generator (http://www.wavtones.com/functiongenerator.php) at a sample rate of 192 kHz and 32-bit resolution. These were played through Beyerdynamic DT 770 PRO Studio Headphones (80 Ohm).

2.2.2. Wines
The wines were two commercially available red wines, selected as they both possessed a similar medium body, but differing levels of acidity and aromas. Wine 1 was Torres Sangre de Toro 2015: 13.5% abv; pH: 3.82; total acidity: 4.9 g/l; residual sugar: 1.2 g/l (Note 7) and with a professional organoleptic assessment suggesting that it possessed a medium (−) aromatic intensity (with subtle notes of ripe plum), medium (−) acidity, medium body and medium (−) tannins. Quality Level: Acceptable — a simple fruity wine with light-to-medium intensity, soft red fruit and spice typical of the variety, fairly short and simple; not suitable for ageing (Note 8). Wine 2 was a Brancott Estate Letter Series 'T' Marlborough Pinot Noir 2015: 14.1% abv; pH: 3.50; total acidity: 5.7 g/l; residual sugar: 2.7 g/l (Note 9) and with a professional organoleptic assessment reporting that it possessed a medium aromatic intensity (with notes of cherry, cinnamon, spice, and herb), medium (+) acidity, medium body and medium tannins. Quality Level: Good — a medium intensity wine with good freshness and brightness to its ripe fruit, showing clear varietal typicity and some depth and complexity; suitable for mid-term ageing. These were served in six 20-ml measures to participants.

2.3. Experimental Design

The aim of the present study was to determine whether the perception of body (along with the two additional wine characteristics: aromatic intensity and acidity) would be altered by tasting a wine while listening to a relatively low-frequency sound (100 Hz), as compared to when tasting in silence and, by way of contrast, a relatively higher frequency sound (1000 Hz). The participants were instructed to rate levels of body, acidity (sourness), aromatic intensity, and liking, in the different sound conditions. Acidity was included in order to offer a point of comparison with previous studies that have robustly demonstrated associations between high acidity and higher pitched sounds in other beverages (Bronner *et al.*, 2012; Crisinel and Spence, 2009, 2010a, b; Knöferle *et al.*, 2015; Kontukosi *et al.*, 2015; Mesz *et al.*, 2011, 2012). Aroma intensity was selected as an extra parameter in order to permit the observation of any olfactory change in an actual wine, given the under-researched nature of this area and potential interaction between aromas and body as noted earlier. The participants rated how much they liked the wine in the different conditions in order to determine whether hedonic factors played any role in determining the participants' responses.

2.4. Experimental Procedure

The experimental procedure comprised two separate studies. The main study is covered here in this section (see Fig. 1a). The supplementary (RATA) study is detailed in section 2.5 (see Fig. 1b).

The main experiment in both locations was conducted in an identical manner using the same wines and sound stimuli. This was performed in a quiet room with one participant at a time, who was seated in front of a computer screen with a pair of headphones, computer mouse, spittoon and crackers. They completed the electronic survey that had been programmed on the Qualtrics online survey platform. Given the difficulty that inexperienced tasters have in identifying body/viscosity in particular (Niimi *et al.*, 2017), participants were first given a brief training session on the wine characteristics that they were to rate. They were provided with a small selection of training wines to smell and taste that illustrated a higher and a medium rating on the nine-point scale they went on to use (see Fig. 2), as well as water, which was used as the example of the lowest rating for aroma, acidity, and body.

The study wines were then presented in a randomized order in black glasses in each sound condition in order to remove any potential visual bias and ensure that each of the wines was evaluated as a fresh sample by the participants in the different sound conditions. Each participant tasted each of the two wines in a silent condition and with both two-minute-length low and higher sine tones, which were also presented in a random order. For each condition, the

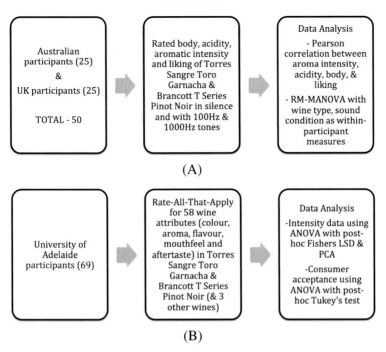

(A)

(B)

Figure 1. (A) Illustration of the methodology applied in the main (pitch-based) study. (B) Illustration of the methodology applied in the RATA study.

participants rated their perception of the level of the wine's aroma, acidity, and body, as well as their liking for the wine, on a nine-point scale (see Fig. 2). The questions were presented as an electronic form using Qualtrics Online Survey Software, with the questions on body and acidity randomized following the question on aroma. They were also asked to rate how much they liked the wine in both silence and the different tone conditions. The participants were asked to answer all the questions while the two-minute tone was presented, with a break of approximately one minute between samples in which participants were advised to cleanse their palates with the water provided.

2.5. Rate-All-That-Apply (RATA)

As body is an element of wine whose overall physical dimensions cannot currently be measured objectively, and for which the judgment of levels re-lies on sensory assessment, the wines used were additionally assessed using a Rate-All-That-Apply (RATA) study in order to obtain additional objective, sensory profiles of the wines that were tasted in the main experiment. For a review of the oral-somatosensory attributes of food and drink, see Spence and Piqueras-Fiszman (2016) and for mouthfeel in general, see Mouritsen and Styrbæk (2017). Further specific overviews are provided for beverages

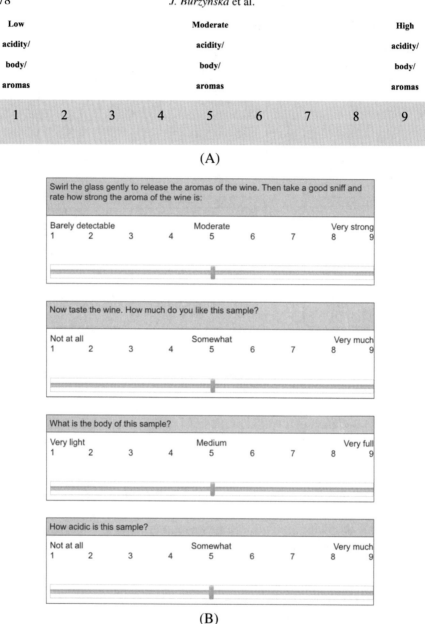

Figure 2. (A) Example of the 9-point scales used for participant ratings; (B) Example of the questionnaire presented on a computer as seen by the participants.

by Szczesniak (1979), and wine by Gawel (2006) and Gawel *et al.* (2018). RATA is an intensity-based variant of Check-All-That-Apply (CATA) developed for sensory characterization using untrained panelists (see Danner *et al.*,

2018). RATA assessments of the Torres Sangre de Toro 2015 and Brancott Estate Letter Series 'T' Marlborough Pinot Noir 2015 were made within six months of the main experiment (Note 10). The RATA analysis was approved by the University of Adelaide Human Research Ethics Committee H-2016-194 and undertaken by 69 panelists consisting of a mix of staff and students from the University of Adelaide's viticulture and oenology programmes between 20 and 56 years of age (27 male, 42 female), who possessed some form of wine evaluation training. The study was conducted in individual, computerised sensory booths, using a methodology identical to that outlined in Danner *et al.* (2017), but covered a range of 58 wine attributes — across colour, aroma, flavour, mouthfeel and aftertaste (see Appendix) and liking.

2.6. Data Analysis

First, Pearson correlation coefficients were calculated between the measures of aroma intensity, acidity, body, and liking. Next, a repeated-measures multivariate analysis of variance (RM-MANOVA) was conducted with wine type (Garnacha, Pinot Noir) and sound condition (silent, 100 Hz, 1000 Hz) as within-participant measures (SPSS, version 23.0, IMB Corp., Armonk, NY, USA). The model included body, acidity, aromatic intensity, and liking as measures (dependent variables). Furthermore, wine familiarity (taken as an average of wine expertise and wine drinking frequency), was introduced as a between-participants factor. Wine familiarity was calculated by taking the average of wine expertise and wine drinking frequency values, and splitting the population into those with low familiarity ($n = 21$) and high familiarity ($n = 29$) (see Fig. 3). Follow-up univariate ANOVAs were conducted on dependent variables where there was a significant main effect or interaction effect amongst the independent variables.

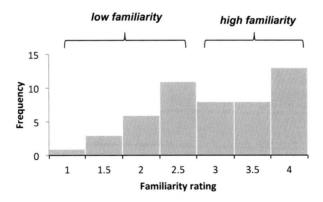

Figure 3. Histogram of participants' self-reported wine familiarity rating. Participants were divided into two equal groups, those with low familiarity (with rating < 3, $n = 21$) and those with high familiarity (with rating ≥ 3, $n = 29$).

RATA intensity data were analysed using two-way Analyses of Variance (ANOVA) (sample as fixed and panelist as random factor), treating the data as continuous data (a non-selected attribute was treated equivalent to 'not perceived' and assigned an intensity of 0), with a post-hoc Fisher's Least Significant Difference (LSD) test and principal component analysis (PCA). Consumer acceptance data were analysed using ANOVA with a post-hoc Tukey's test. Data were analysed using Senpaq v5.01 (Qi Statistics, Theale, UK) and XLSTAT Version 2016.03.31333 (Addinsoft, New York, NY, USA). All statistical analyses were performed at 5% level of significance.

3. Results

Significant correlations were found between ratings of aroma intensity, acidity, body, and liking in the main study (see Table 1). Notably, all pairwise correlations were positive. For instance, wine liking was positively correlated with aroma intensity, acidity, and body. Moreover, perceived body was positively correlated with aromatic intensity and acidity.

The mean values of the participants' wine ratings for both types of wine and all three sound conditions are shown in Fig. 4. Overall, the RM-MANOVA revealed a significant main effect of wine type [$F(4, 45) = 9.21$, $p < 0.0005$, Wilks' Lambda $= 0.55$], as well as a significant interaction effect between wine type and sound condition [$F(8, 41) = 2.36$, $p = 0.034$, Wilks' Lambda $= 0.68$]. However, there was no significant main effect of sound condition [$F(8, 41) = 1.70$, $p = 0.13$], nor of wine familiarity [$F(4, 45) = 2.38$, $p = 0.07$] on the dependent measures (body, acidity, aromatic intensity, and liking). No significant differences were noted between the responses of the Sydney and Oxford participant groups.

In terms of wine type, follow-up univariate tests revealed that there were significant differences between the two wines in terms of their aroma intensity [$F(1, 48) = 24.42$, $p < 0.0005$, $\eta^2 = 0.34$] and participants' liking ([$F(1, 48) = 6.72$, $p = 0.013$, $\eta^2 = 0.12$]. Overall, the Pinot Noir was rated

Table 1.

Pearson correlation coefficients ($n = 189$) amongst participants' ratings of aroma intensity, acidity, body, and liking. * indicates significance at <0.05 level, ** indicates significance at <0.01 level

	Aroma intensity	Acidity	Body	Liking
Aroma intensity	1	0.21 **	0.21 **	0.20 **
Acidity		1	0.23 **	0.14 *
Body			1	0.43 **
Liking				1

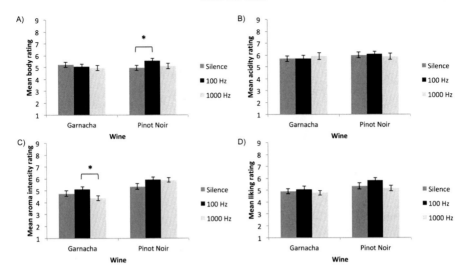

Figure 4. Mean ratings of body (A), acidity (B), aroma intensity (C), and liking (D) for both wines in the three sound conditions: Silence, low pitch (100 Hz), and high pitch (1000 Hz). Error bars indicate standard error. Asterisks indicate statistical significance at $p < 0.05$.

as more aromatic than the Garnacha [M_{Pinot} (SD) = 5.68 (1.82), $M_{Garnacha}$ (SD) = 4.69 (1.51), $p < 0.0005$]. In terms of liking, the Pinot Noir was liked significantly more than the Garnacha [M_{Pinot} (SD) = 5.38 (1.68), $M_{Garnacha}$ (SD) = 4.90 (1.44), $p = 0.013$].

Furthermore, there were significant interaction effects between wine type and sound condition for the ratings of aroma intensity [$F(2, 96) = 3.48$, $p = 0.035$, $\eta^2 = 0.07$] and body [$F(2, 96) = 3.84$, $p = 0.025$, $\eta^2 = 0.074$]. For aroma intensity, the interaction was driven by the fact that the Garnacha was rated to be significantly more aromatically intense while participants were listening to the 100 Hz low tone rather than the 1000 Hz higher tone [M_{100Hz} (SD) = 5.14 (1.64), M_{1000Hz} (SD) = 4.38 (1.43), $p = 0.007$]. No such differences were found for the Pinot Noir. In terms of body, wines were rated as being significantly fuller while listening to the 100 Hz low tone as compared to silence for the Pinot Noir [M_{100Hz} (SD) = 5.54 (1.64), $M_{Silence}$ (SD) = 4.96 (1.59), $p = 0.021$]. However, the same effect was not observed for the Garnacha, in which there was little change between the different conditions.

While there was no main effect of wine familiarity, we did observe a significant interaction effect between sound condition and wine familiarity when it came to ratings of acidity [$F(2, 96) = 3.84$, $p = 0.025$, $\eta^2 = 0.074$]. As Fig. 5 illustrates, this interaction was driven by the difference in acidity ratings in the high frequency (1000 Hz) condition, with those with high wine familiarity perceiving the wines as much more acidic compared to those with low wine familiarity [M_{Low} (SD) = 5.21 (1.57), M_{High} (SD) = 6.40 (1.61), $p = 0.004$].

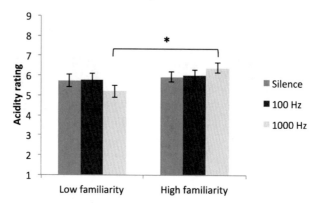

Figure 5. Mean ratings of acidity in the three sound conditions: silence, low pitch (100 Hz), high pitch (1000 Hz), grouped by wine familiarity (low, $n = 21$ or high, $n = 29$). Error bars indicate standard error. Asterisks indicate statistical significance at $p < 0.05$.

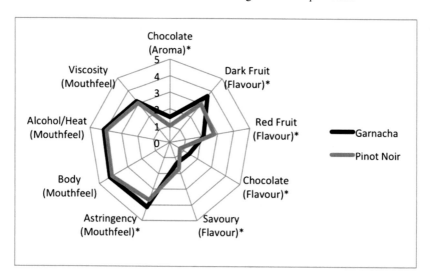

Figure 6. Mean intensity ratings of selected attributes from RATA analysis rated on a scale of 1–7. Those differences that are statistically significant (at $p < 0.05$) are marked with an asterisk.

Non-significant trends suggested that those with high wine familiarity rated the wines to be the *most* acidic in the 1000 Hz condition out of all sound conditions, whereas those with low wine familiarity rated the wines to be the *least* acidic in the 1000 Hz condition out of all sound conditions.

The RATA analysis revealed that the Garnacha and Pinot Noir wines were perceived as similar in most attributes. Relevant to the main study, this included comparable perceptions of acidity, body and viscosity in both wines (see Fig. 6). However, there were six significant differences ($p < 0.05$) discovered. The Garnacha was rated higher in chocolate aroma, chocolate and dark fruit flavours, and astringency, while the Pinot Noir was rated higher in red fruit and savoury notes (see Fig. 4). Liking was similar across all six wines in the RATA study, with the mean liking rating of the Garnacha 5.25 and the Pinot Noir 5.12 on a nine-point hedonic scale. A further visual examination of a biplot of the Principal Component Analysis, which explained 82.08% of the variation of the data in the sensory space in the first two components (data not shown), of all the six wines in the RATA study showed the Garnacha positioned in the sensory space towards the left of the plot, which contained attributes such as confectionary and jammy flavours, while the Pinot Noir was situated to the right, closer to savoury and earthy/dusty flavours.

These results demonstrate a significant cross-modal interaction between sound conditions (silence, low tone, or high tone) and wine type (a Spanish Garnacha and a New Zealand Pinot Noir). In the case of the Garnacha, this wine's aromatic intensity was perceived as being significantly higher when tasted while listening to the low tone than when tasted while listening to the high tone. In the case of the NZ Pinot Noir, the body of the wine was rated on average as being significantly fuller while listening to the low-pitched tone than when tasted in silence. However, there was no effect of the low-pitched tone on the perception of body for the Garnacha nor on the perception of the aromatic intensity of the Pinot Noir.

4. Discussion and Conclusions

In the study reported here, listening to a bass tone (100 Hz) was shown to elicit significant changes in the perception of characters of both red wines tested. This pattern of results suggests that pitch not only affects the perception of basic tastes, but also of aromatics and the mouthfeel character of body in a wine as well. As suggested by widely held associations and metaphorical connections between weight and low pitch, in the case of the NZ Pinot Noir, a low frequency sound significantly increased perceptions of body in a wine. Bass also created a significant augmentation in the perception of aromatic intensity in the Spanish Garnacha. However, as to the question of why the shifts in perception of these attributes should have been different for the two wines the answer is currently less clear.

When looking for possible reasons behind the difference in aromatic augmentation between the two wines, it should be noted that the Garnacha was rated as the lower of the two wines in terms of its aromatic intensity in the silent condition and in the professional organoleptic assessment. It could therefore be proposed that in possessing less aromatic intensity there would be more scope for movement within aromatic ratings for this wine as compared with that of the NZ Pinot Noir, which was a more aromatic wine to begin with.

Given previous research examining associations between pitch and aroma ascertained matches between different aromas and different pitches (see Belkin et al., 1997; Crisinel and Spence, 2011), it could be that the different aromatic characters found in the wines could have had an effect on perceived intensity. The RATA study identified that the aromatic profile of the Garnacha differed significantly from that of the Pinot Noir in possessing higher levels of chocolate aromas. Crisinel and Spence's study, which used samples from the *Nez du Vin* wine aroma kit as olfactory stimuli, found chocolate aromas were associated with low-pitched notes. From this, it could be surmised that presence of chocolate aromas in the Garnacha, in being congruent with low frequencies, caused the aromatic intensity to be perceived as greater with the bass tone in this particular wine.

While the wines possessed different levels of aromatic intensity, their body was comparable in both the silent condition of the study, the professional assessment, and the RATA study. This makes the increase in perceived body in the low pitch condition for the Pinot Noir, but not the Garnacha (where very little change in all conditions was apparent), harder to unravel. Noted as an "essential factor" in the choice of pitch in Crisinel and Spence's (2012) study of cross-modal associations between musical notes and odours, complexity could be another element promoting the different responses to a similarly weighted attribute (body) in this study. As noted by Lavie (2005), more complex perceptual stimuli increase the effect of attention on awareness, although effects are somewhat mixed for audition (Murphy *et al.*, 2017). The various assessments of the Garnacha reveal it to be less complex than the Pinot Noir. It was judged to be 'a simple fruity wine' of 'acceptable' quality in the professional organoleptic assessment, and located closer to the area of simple ('jammy' and 'confectionary') fruit flavours on the RATA biplot. In contrast, the Pinot Noir was identified in the professional assessment as possessing some complexity and judged as 'good', while in the RATA results it was situated closer to what could be considered more complex 'savoury' and 'earthy/dusty' flavours on the biplot. This could have lessened attention to body in the case of the Garnacha and increased it in the case of the Pinot Noir.

Emotion has been put forward as a mechanism mediating cross-modal correspondences between taste and sound (Wang *et al.*, 2016), odour and pitch (Crisinel and Spence, 2012), and taste and shape (Turoman *et al.*, 2018; Velasco *et al.*, 2015). It could be surmised that emotional factors (what is known as emotional mediation) may have played a role here, given that there was a difference in the liking of the two wines. The Pinot Noir, which elicited responses to the bass frequency in the manner hypothesized, was the significantly better-liked wine of the two. This suggests that hedonic mediation might be a factor in the different responses related to the interaction between bass and body with the two wines. However, it should be noted that there were no significant differences in hedonic ratings of the wines between the different sound conditions. As no emotional values were collected for the two auditory pitches used in this experiment — only the wines were rated for liking — the possibility of emotional mediation cannot be extrapolated from within this experiment itself. Some illumination, however, could be provided by cross referencing the pitches used in our study with those investigated in another study by Wang *et al.* (2016) that investigated the role of emotion in mediating correspondences between basic tastes and sound (Note 11). This suggests that while mediation might not have been hedonic in our study, arousal could have been a factor mediating the correspondence between low pitch and full body in the Pinot Noir in our study. The Pinot Noir could feasibly be regarded as the more exciting wine in both being better liked, and more complex than the Garnacha. In light of this, in future experiments it might be advisable if more emotional dimensions are included in relation to the sounds, as well as the wines.

While the complexity of the wine (see Spence and Wang, 2018, on the challenging topic of wine complexity) offered the potential for multiple correspondences, its complex multidimensional nature also makes it a challenging stimulus to assess. Given the diversity and complexity of the wine chemistry and the perception of its sensory characters in both wines in this study, there could have been a number of attributes that influenced each other, and consequently on the way their flavour (and aroma) profiles were perceived when accompanied with the sounds. As greater viscosity tends to be associated with a lessened perception of intensity of flavours and volatile components in model solutions, depending on the compounds (Tournier *et al.*, 2007; see Spence and Piqueras-Fiszman, 2016, for a review), that there were significant pairwise correlations between ratings of aroma intensity, acidity, body, and liking, could suggest that people might be transferring high intensity across attributes. This could also be due to learned associations that link high levels of attributes, as demonstrated by novice wine drinkers associating fuller bodied wines with greater flavor intensity and vice versa (Niimi *et al.*, 2017). However, no research examining these specific taste–aroma interactions in wine has been published to date. Furthermore, it should be noted that as the area of taste-aroma interactions is a complex one (Paravisini and Guichard, 2017; Tournier *et al.*, 2007) and study into cross-modal correspondences between sound and wine is still at a nascent stage, more research is required into how the different attributes of both wine characters, and wine and sound interact.

When the data from the participants was segmented into groups, correlations between higher acidity and higher pitch emerged as a trend in the group with greater wine familiarity. This is a mapping demonstrated in numerous other studies that might therefore have been expected in this one. A recent study by Wang and Spence (2018) indicated cross-modal effects between music and wine were similar across experts and non-experts. However, its authors also proposed that experts may be better placed to detect the subtle attentional effects of music on their taste perception, which one could surmise might have been the case in our study. Why the sour–pitch correspondence did not emerge so clearly in this study could also — as previously discussed — potentially be due to the complexity of wine, in which acidity is just one of many salient perceptual attributes. It can also be noted that a 1000 Hz tone is not particularly high.

From looking at these different participant categories, it can also be noted that the more experienced tasters performed in a far more consistent manner in their judgments than did the beginners. This could suggest that the methodology of the experiment was better suited to those with more wine experience. It could be hypothesised that experts might be more used to conceptualizing levels of the attributes selected in the study, rating these on a scale. This was perhaps more of a challenge to novices (see discussion of rating scales by Meilgaard *et al.*, 2006, pp. 55–60). Regular wine drinkers could also be more confident in their interactions with wine, given that it is a regular part of their lifestyle, in contrast with those for whom wine constituted an occasional drink. It should be noted that some less experienced tasters among participants voiced concern to the invigilator with regards to their accurate use of the scale. It could be that any changes needed to be more overt for them to be rated more decisively, in which case such a strong result for the low frequency increasing perceptions of the Pinot Noir's body suggests that this is particularly strong cross-modal correspondence.

Issues with rating on scales could be overcome, and the results clarified, by running an additional study using a different methodology in order to further explore possible correlations between bass and body through, for example, a matching exercise. In this case, the participants would be given three wines of varying body (light, medium, and full), but with their other main parameters (i.e., acidity, tannins) as similar as possible, to match with three tones of varying pitch (low, medium, and high). In this simple matching exercise any correlations between pitch and body should become more apparent.

When interpreting the data, it is also worth considering the relationship between pitch and body in light of current discussions regarding the relative versus absolute nature of the cross-modal correspondences (see Brunetti *et al.*, 2018; and Spence, 2019c). Past research suggests that most cross-modal correspondences involving the metathetic pitch dimension are relative (Ben-Artzi and Marks, 1995; Brunetti *et al.*, 2018; Chiou and Rich, 2012; Stevens, 1957). The range of responses in this study suggests that the pitch–body and pitch–aroma correspondences identified are not absolute. The completely randomized presentation of the tones with the two wines in this study could mean that for some participants the same, rather than a contrasting, tone was presented sequentially. The first tone presented might also have been difficult to categorize as being either high or low. Brunetti *et al.*'s (2018) investigation highlighted the flexibility and sequential influence of stimuli presentation in relation to the absolute versus relative nature of pitch-size correspondences. From this, it could be deduced that the randomization in our study could have lessened the correspondences discovered. To overcome these possible effects, future experiments could expose participants to the range of tones used before the start of the experiment. Using a tone higher in frequency than 1000 Hz might presumably also assist in greater differentiation between the tones by participants.

The perception of low pitch would appear, in some wines at least, to significantly increase the perception of body. This finding highlights that this approach should be a fruitful avenue for further research to clarify the prevalence of this correspondence in wine, as well as in other beverages and foodstuffs. That bass notes were able to shift the perception of aromatics in one wine (the Garnacha) and increase the perception of the attribute of body in another (the Pinot Noir), provides a step towards the greater understanding of the web of relationships between the complex flavour profiles of wines and the influence of sounds on their perception.

These findings could be used to inform the wine and sound/music combinations created by those working with wine and sound creatively — such as artists, musicians, designers and chefs. This could permit greater control in the shaping of multisensory environments through using more targeted "sonic seasoning" (Spence, 2017). In bass-heavy environments, such as clubs and bars, this knowledge could be of particular significance in highlighting that the music played will likely be having a significant impact on the taste experiences of their clientele. Conversely, when people are making their beverage selections in these clubs, they might use these observations to choose a drink that improves with the perception of increased body, such as a fuller bodied red wine, rather than one that is adversely affected by it, for example, a Champagne or light white wine. These findings could also be of value for consideration in the acoustic design of spaces where fine wine is to be consumed, suggesting attention needs to be paid to controlling bass frequencies through the use of appropriate low frequency sound-absorbing materials.

Given the growing trend of playing music at both wine tastings and even during the judging of professional wine competitions, these cross-modal correspondences between flavour/aroma and sound may also be applicable for consideration by those organizing such events. Music is played at competitions during the judging process, at competitions such as the UK's International Wine Competition (IWC) and Australia's Adelaide Review Hot 100.

This is despite ISO 8589 (2007) guidelines for the design of test rooms intended for the sensory analysis of products, which advises in such spaces "noise level shall be kept to a minimum". These findings reinforce the need for quiet environments to minimize not only the possibility for distraction, but sonic influence on the perception of flavour and aroma characters during wine assessment. This would appear particularly relevant to professional tastings, where a consistent evaluation of the wines is desired. Conversely, in more entertainment-based consumer wine events, where music can play positive roles in priming moods and creating an appropriate ambience, these correspondences can be applied to the sound mix to emphasize desired wine characters.

Acknowledgements

The authors wish to thank Dr Trent Johnson of the University of Adelaide for facilitating the RATA panel and data analyses. Jo Burzynska's research was supported through an Australian Government Research Training Program Scholarship. Wines supplied by Pernod Ricard and Bodegas Torres.

Notes

1. This legend is noted in Zbikowski (2002).

2. Text from a DMZ (Digital Mystikz) dubstep club invitation referenced in Jasen (2017). Dubstep is a bass-heavy genre of electronic dance music.

3. Oenosonic is a term coined by Jo Burzynska (2018b) to describe the creative combination of wine and sound in her works.

4. The wine was the Crown Range Grant Taylor Signature Series Central Otago Pinot 2016.

5. "The heavy bass of the Dubstep music ... enhanced the body of the wine and our perception of umami." Sara d'Amato, sommelier. https://www.winealign.com/articles/2017/03/02/buyers-guide-to-vintages-march-4th-2017/ Accessed 5 November 2018.

6. As noted by Leventhall *et al.* (2003), it would appear that some people are sensitive to low frequencies (i.e., 10 Hz to 200 Hz) and experience distress in their presence.

7. Technical specifications provided by Bodegas Torres.

8. Assessed by Jo Burzynska, a professional wine judge and Certified Wine and Spirit Trust (WSET) Educator and holder of a WSET Diploma, using the WSET Systematic Approach to Tasting Wine®.

9. Technical specifications provided by Pernod-Ricard.

10. As a minimum of six wines are required for multivariate analyses, this study also included another cool climate Pinot Noir from the Adelaide Hills wine region, South Australia and three warm climate Garnachas/Grenaches from McLaren Vale, South Australia which were not examined further in the main study.

11. Wang *et al.* (2016) used the dimensions of valence and arousal, which were mapped by the study's participants to pitch (using a piano sound). Cross-referencing the closest pitches used in our experiment with those of Wang *et al.*'s (2016: Appendix B) study, the lower pitch note was less liked than the higher pitch note, which casts some doubt over hedonic mediation in our study. However, in terms of arousal, the lower pitch was rated as more arousing than the higher pitch.

References

Belkin, K., Martin, R., Kemp, S. E. and Gilbert, A. N. (1997). Auditory pitch as a perceptual analogue to odor quality, *Psychol. Sci.* **8**, 340–342.

Ben-Artzi, E. and Marks, L. E. (1995). Visual-auditory interaction in speeded classification: role of stimulus difference, *Percept. Psychophys.* **57**, 1151–1162.

Bronner, K., Bruhn, H., Hirt, R. and Piper, D. (2012). What is the sound of citrus? Research on the correspondences between the perception of sound and flavour, in: *Proceedings of the 12th International Conference of Music Perception and Cognition (ICMPC) and the 8th Triennial Conference of the European Society for the Cognitive Sciences of Music (ESCOM), Thessaloniki, Greece*, pp. 142–148.

Brunetti, R., Indraccolo, A., Del Gatto, C., Spence, C. and Santangelo, V. (2018). Are crossmodal correspondences relative or absolute? Sequential effects on speeded classification, *Atten. Percept. Psychophys.* **80**, 527–534.

Burzynska, J. (2018a). Assessing *oenosthesia*: blending wine and sound, *Int. J. Food Des.* **3**, 83–101.

Burzynska, J. (2018b). Oenosonic immersion at Pollinaria. http://joburzynska.com/oenosonic-immersion-at-pollinaria/. Retrieved 14 March 2019.

Chiou, R. and Rich, A. N. (2012). Cross-modality correspondence between pitch and spatial location modulates attentional orienting, *Perception* **41**, 339–353.

Crisinel, A.-S. and Spence, C. (2009). Implicit association between basic tastes and pitch, *Neurosci. Lett.* **464**, 39–42.

Crisinel, A.-S. and Spence, C. (2010a). As bitter as a trombone: synesthetic correspondences in nonsynesthetes between tastes/flavors and musical notes, *Atten. Percept. Psychophys.* **72**, 1994–2002.

Crisinel, A.-S. and Spence, C. (2010b). A sweet sound? Food names reveal implicit associations between taste and pitch, *Perception* **39**, 417–425.

Crisinel, A.-S. and Spence, C. (2011). A fruity note: crossmodal associations between odors and musical notes, *Chem. Sens.* **37**, 151–158.

Crisinel, A.-S. and Spence, C. (2012). The impact of pleasantness ratings on crossmodal associations between food samples and musical notes, *Food Qual. Pref.* **24**, 136–140.

Crisinel, A.-S., Cosser, S., King, S., Jones, R., Petrie, J. and Spence, C. (2012). A bittersweet symphony: systematically modulating the taste of food by changing the sonic properties of the soundtrack playing in the background, *Food Qual. Pref.* **24**, 201–204.

D'Amato, S. (2017). Wine in the sky, edgy finds and New World gems. https://www.winealign. com/articles/2017/03/02/buyers-guide-to-vintages-march-4th-2017/. Retrieved 5 November 2018.

Danner, L., Crump, A. M., Croker, A., Gambetta, J. M., Johnson, T. E. and Bastian, S. E. P. (2018). Comparison of rate-all-that-apply and descriptive analysis for the sensory profiling of wine, *Am. J. Enol. Viticult.* **69**, 12–21.

Datsik (2009). Gizmo / Gecko, Basshead. Mp3.

Deroy, O., Crisinel, A.-S. and Spence, C. (2013). Crossmodal correspondences between odors and contingent features: odors, musical notes, and geometrical shapes, *Psychon. Bull. Rev.* **20**, 878–896.

Deroy, O., Fernandez-Prieto, I., Navarra, J. and Spence, C. (2018). Unravelling the paradox of spatial pitch, in: *Spatial Biases in Perception and Cognition*, T. L. Hubbard (Ed.), pp. 77–93. Cambridge University Press, Cambridge, UK.

Dolscheid, S., Hunnius, S., Casasanto, D. and Majid, A. (2014). Prelinguistic infants are sensitive to space-pitch associations found across cultures, *Psychol. Sci.* **25**, 1256–1261.

Gacula Jr, M. and Rutenbeck, S. (2006). Sample size in consumer test and descriptive analysis, *J. Sens. Stud.* **21**, 129–145.

Galeyev, B. M. (2003). Evolution of gravitational synesthesia in music: to color and light! *Leonardo* **36**, 129–134.

Gawel, R. (2006). Importance of 'texture' to red wine quality acknowledged by the development of a red wine "Mouth-feel Wheel". http://www.aromadictionary.com/articles/ winemouthfeel_article.html. Retrieved 1 February 2006.

Gawel, R., Schulkin, A., Smith, P., Kassara, S., Francis, L., Herderich, M. and Johnson, D. (2018). Influence of wine polysaccharides on white and red wine mouthfeel, *Wine Viticult. J.* **33**, 34–37.

Grainger, K. (2009). *Wine Quality: Tasting and Selection*. John Wiley and Sons, Oxford, UK.

Henriques, J. (2003). Sonic dominance and the reggae sound system session, in: *The Auditory Culture Reader*, M. Bull and L. Black (Eds), pp. 451–480. Berg, Oxford, UK.

Henriques, J. (2011). *Sonic Bodies: Reggae Sound Systems, Performance Techniques, and Ways of Knowing*. Bloomsbury, New York, NY.

Holt-Hansen, K. (1968). Taste and pitch, *Percept. Mot. Skills* **27**, 59–68.

Holt-Hansen, K. (1976). Extraordinary experiences during cross-modal perception, *Percept. Mot. Skills* **43**, 1023–1027.

Huysmans, J.-K. (1959). *Against Nature* [a new translation of À Rebours]. Transl. by Robert Baldick. Penguin Books, London, UK.

International Organization for Standardization (2003). *Acoustics — Normal equal-loudness-level contours*. (ISO Standard No. 226). https://www.iso.org/standard/34222.html. Retrieved 4 October 2017.

International Organization for Standardization (2007). *Sensory analysis — General guidance for the design of test rooms*. (ISO Standard No 8589). https://www.iso.org/standard/36385. html. Retrieved 5 November 2018.

Jackson, R. S. (2009). *Wine Tasting: a Professional Handbook*. Academic Press, London, UK.

Jasen, P. C. (2017). *Low End Theory: Bass, Bodies and the Materiality of Sonic Experience*. Bloomsbury Press, London, UK.

Knoeferle, K. M., Woods, A., Käppler, F. and Spence, C. (2015). That sounds sweet: using cross-modal correspondences to communicate gustatory attributes, *Psychol. Market.* **32**, 107–120.

Knöferle, K. and Spence, C. (2012). Crossmodal correspondences between sounds and tastes, *Psychon. Bull. Rev.* **19**, 992–1006.

Kontukoski, M., Luomala, H., Mesz, B., Sigman, M., Trevisan, M., Rotola-Pukkila, M. and Hopia, A. I. (2015). Sweet and sour: music and taste associations, *Nutr. Food Sci.* **45**, 357–376.

Laguna, L., Bartolomé, B. and Moreno-Arribas, M. V. (2017). Mouthfeel perception of wine: oral physiology, components and instrumental characterization, *Trends Food Sci. Technol.* **59**, 49–59.

Lavie, N. (2005). Distracted and confused?: selective attention under load, *Trends Cogn. Sci.* **9**, 75–82.

Leventhall, G., Pelmear, P. and Benton, S. (2003). *A review of published research on low frequency noise and its effects*. Department for Environment, Food and Rural Affairs, UK. https://westminsterresearch.westminster.ac.uk/item/935y3/a-review-of-published-research-on-low-frequency-noise-and-its-effects. Retrieved 29 April 2019.

Macdermott, M. M. (1940). *Vowel Sounds in Poetry: Their Music and Tone-Color*. Kegan Paul, London, UK.

McAngus Todd, N. P., Rosengren, S. M. and Colebatch, J. G. (2008). Tuning and sensitivity of the human vestibular system to low-frequency vibration, *Neurosci. Lett.* **444**, 36–41.

Meilgaard, M. C., Civille, G. V. and Carr, B. T. (2006). *Sensory Evaluation Techniques*. CRC Press, Cleveland, OH, USA.

Mesz, B., Trevisan, M. A. and Sigman, M. (2011). The taste of music, *Perception* **40**, 209–219.

Mesz, B., Sigman, M. and Trevisan, M. A. (2012). A composition algorithm based on cross-modal taste-music correspondences, *Front, Hum. Neurosci.* **6**, 71. DOI:10.3389/fnhum.2012.00071.

Moran, M. (2017). Tannins are all about that bass. *Sunday Times*, London, UK, 26 Feb, 13.

Mouritsen, O. and Styrbæk, K. (2017). *Mouthfeel: How Texture Makes Taste* (transl. M. Johansen). Columbia University Press, New York, NY, USA.

Murphy, S., Spence, C. and Dalton, P. (2017). Auditory perceptual load: a review, *Hear. Res.* **352**, 40–48.

Niimi, J., Danner, L., Li, L., Bossan, H. and Bastian, S. E. P. (2017). Wine consumers' subjective responses to wine mouthfeel and understanding of wine body, *Food Res. Int.* **99**, 115–122.

Oxford Living Dictionaries. Retrieved 5 November 2018. https://en.oxforddictionaries.com/definition/heaviness.

Paravisini, L. and Guichard, E. (2017). Interactions between aroma compounds and food matrix, in: *Flavour: from Food to Perception*, E. Guichard, C. Salles, M. Morzel and A.-M. Le Bon (Eds), pp. 208–234. Wiley-Blackwell, Chichester, UK.

Parise, C. V. (2016). Crossmodal correspondences: standing issues and experimental guidelines, *Multisens. Res.* **29**, 7–28.

Parise, C. V., Knorre, K. and Ernst, M. O. (2014). Natural auditory scene statistics shapes human spatial hearing, *Proc. Natl Acad. Sci. USA* **111**, 6104–6108.

Reinoso-Carvalho, F., Wang, Q. (J.), De Causmaecker, B., Steenhaut, K., Van Ee, R. and Spence, C. (2016a). Tune that beer! Listening for the pitch of beer, *Beverages* **2**, 31. DOI:10.3390/beverages2040031.

Reinoso-Carvalho, F., Wang, Q. (J.), Van Ee, R. and Spence, C. (2016b). The influence of soundscapes on the perception and evaluation of beers, *Food Qual. Pref.* **52**, 32–41.

Rudmin, F. and Cappelli, M. (1983). Tone–taste synesthesia: a replication, *Percept. Motor Skills* **56**, 118. DOI:10.2466/pms.1983.56.1.118.

Shayan, S., Ozturk, O. and Sicoli, M. A. (2011). The thickness of pitch: crossmodal metaphors in Farsi, Turkish, and Zapotec, *Sens. Soc.* **6**, 96–105.

Simner, J., Cuskley, C. and Kirby, S. (2010). What sound does that taste? Cross-modal mapping across gustation and audition, *Perception* **39**, 553–569.

Spence, C. (2011). Wine and music, *World Fine Wine* **31**, 96–104.

Spence, C. (2014). Noise and its impact on the perception of food and drink, *Flavour* **3**, 9. DOI:10.1186/2044-7248-3-9.

Spence, C. (2015). Eating with our ears: assessing the importance of the sounds of consumption on our perception and enjoyment of multisensory flavour experiences, *Flavour* **4**, 3. DOI:10.1186/2044-7248-4-3.

Spence, C. (2017). Sonic seasoning, in: *Audio Branding: Using Sound to Build Your Brand*, L. Minsky and C. Fahey (Eds), pp. 52–58. Kogan Page, London, UK.

Spence, C. (2019a). Multisensory experiential wine marketing, *Food Qual. Pref.* **71**, 106–116.

Spence, C. (2019b). On the relationship(s) between colour and taste/flavor, *Exp. Psychol.* **66**, 99–111. DOI:10.1027/1618-3169/a000439.

Spence, C. (2019c). On the relative nature of (pitch-based) crossmodal correspondences, *Multisens. Res.* **32**, 235–265. DOI:10.1163/22134808-20191407.

Spence, C. and Piqueras-Fiszman, B. (2016). Oral-somatosensory contributions to flavor perception and the appreciation of food and drink, in: *Multisensory Flavor Perception: from Fundamental Neuroscience Through to the Marketplace*, B. Piqueras-Fiszman and C. Spence (Eds), pp. 59–79. Elsevier/Woodhead Publishing, Duxford, UK.

Spence, C. and Wang, Q. (J.) (2015a). Wine and music (I): on the crossmodal matching of wine and music, *Flavour* **4**, 34. DOI:10.1186/s13411-015-0045-x.

Spence, C. and Wang, Q. (J.) (2015b). Wine and music (II): can you taste the music? Modulating the experience of wine through music and sound, *Flavour* **4**, 33. DOI:10.1186/s13411-015-0043-z.

Spence, C. and Wang, Q. J. (2015c). Wine and music (III): so what if music influences the taste of the wine? *Flavour* **4**, 36. DOI:10.1186/s13411-015-0046-9.

Spence, C. and Wang, Q. J. (2018). What does the term 'complexity' mean in the world of wine? *Int. J. Gastron. Food Sci.* **14**, 45–54.

Stevens, S. S. (1957). On the psychophysical law, *Psychol., Rev.* **64**, 153–181.

Szczesniak, A. S. (1979). Classification of mouthfeel characteristics of beverages, in: *Food Texture and Rheology*, P. Sherman (Ed.), pp. 1–20. Academic Press, London, UK.

Tournier, C., Sulmont-Rossé, C. and Guichard, E. (2007). Flavour perception: aroma, taste and texture interactions, *Food* **1**, 246–257.

Turoman, N., Velasco, C., Chen, Y.-C., Huang, P.-C. and Spence, C. (2018). Tasting transformations: symmetry and its role in the crossmodal correspondence between shape and taste, *Atten. Percept. Psychophys.* **80**, 738–751.

Velasco, C., Woods, A. T., Deroy, O. and Spence, C. (2015). Hedonic mediation of the cross-modal correspondence between taste and shape, *Food Qual. Pref.* **41**, 151–158.

Walker, L., Walker, P. and Francis, B. (2012). A common scheme for cross-sensory correspondences across stimulus domains, *Perception* **41**, 1186–1192.

Walker, P. (2016). Cross-sensory correspondences: a theoretical framework and their relevance to music, *Psychomusicol. Music Mind Brain* **26**, 103–116.

Walker, P., Scallon, G. and Francis, B. (2017). Cross-sensory correspondences: heaviness is dark and low-pitched, *Perception* **46**, 772–792.

Wang, Q. (J.) and Spence, C. (2015). Assessing the effect of musical congruency on wine tasting in a live performance setting, *i-Perception* **6**. DOI:10.1177/2041669515593027.

Wang, Q. (J.) and Spence, C. (2018). Assessing the influence of music on wine perception among wine professionals, *Food Sci. Nutr.* **6**, 295–301.

Wang, Q. (J.), Woods, A. and Spence, C. (2015). "What's your taste in music?" A comparison of the effectiveness of various soundscapes in evoking specific tastes, *i-Perception* **6**, 2041669515622001. DOI:10.1177/2041669515622001.

Wang, Q. J., Wang, S. and Spence, C. (2016). "Turn up the taste": assessing the role of taste intensity and emotion in mediating crossmodal correspondences between basic tastes and pitch, *Chem. Sens.* **41**, 345–356.

Yan, K. S. and Dando, R. (2015). A crossmodal role for audition in taste perception, *J. Exp. Psychol. Hum. Percept. Perform.* **41**, 590–596.

Zampini, M. and Spence, C. (2004). The role of auditory cues in modulating the perceived crispness and staleness of potato chips, *J. Sens. Stud.* **19**, 347–363.

Zbikowski, L. M. (1998). Metaphor and music theory: reflections from cognitive science, *Music Theory Online* **4**. http://www.mtosmt.org/issues/mto.98.4.1/mto.98.4.1.zbikowski.html. Retrieved 2 October 2017.

Zbikowski, L. M. (2002). *Conceptualizing Music: Cognitive Structure, Theory, and Analysis.* Oxford University Press, New York, NY, USA.

Appendix.
Attributes rated in the RATA study

Colour	Aroma	Flavour by mouth	Mouthfeel	Aftertaste
Red	Dark fruit *blackberry, blackcurrant, plum, dark cherry*	Dark fruit *blackberry, blackcurrant, plum, dark cherry*	Body	Length of fruit flavours
Purple	Red fruit *raspberry, strawberry, red cherry, redcurrant*	Red fruit *raspberry, strawberry, red cherry, redcurrant*	Alcohol/Heat	Length of non-fruit flavours
Brown	Dried fruit *prune, raisin, figs, dates* Jammy *any fruit jam* Confectionary Chocolate	Dried fruit *prune, raisin, figs, dates* Jammy *any fruit jam* Confectionary Chocolate	Astringency Smoothness Roughness Viscosity *the resistance of the wine when you move it around on the palate*	
	Coconut Cooked vegetables *cooked cabbage and beans* Earthy/Dusty Eucalypt/Mint Floral/Perfume/Musk Forest floor *including mushrooms* Green pepper/Capsicum Herbaceaous Leather Pepper *black or white* Savoury *savoury, meaty and gamey* Spice *anise, clove, cinnamon, licorice, nutmeg* Stemmy/Stalky Toasty/Smoky	Coconut Cooked vegetables *cooked cabbage and beans* Earthy/Dusty Evolved/Mature Floral Forest floor *including mushrooms* Green pepper/Capsicum Herbaceaous Leafy Pepper *black or white* Savoury *savoury, meaty and gamey* Spice *anise, clove, cinnamon, licorice, nutmeg* Stemmy/Stalky Toasty/Smoky Vanilla Woody *cedar, pencil shavings, cigar box, tobacco*		

Text in italics indicates description given to panelists for that particular attribute.

Analysing the Impact of Music on the Perception of Red Wine via Temporal Dominance of Sensations

Qian Janice Wang [1,2,*], **Bruno Mesz** [3], **Pablo Riera** [4], **Marcos Trevisan** [5], **Mariano Sigman** [6,7], **Apratim Guha** [8] **and Charles Spence** [1]

[1] Crossmodal Research Laboratory, Department of Experimental Psychology, Oxford University, Oxford, UK

[2] Department of Food Science, Aarhus University, Aarslev, Denmark

[3] MUNTREF Tecnópolis, Universidad Nacional de Tres de Febrero, Buenos Aires, Argentina

[4] Laboratorio de Inteligencia Artificial Aplicada, Instituto de Ciencias de la Computación, Universidad de Buenos Aires, CONICET, Argentina

[5] Department of Physics, University of Buenos Aires and Institute of Physics Buenos Aires (IFIBA), CONICET, Argentina

[6] Laboratorio de Neurociencia, CONICET, Universidad Torcuato Di Tella, C1428BIJ Buenos Aires, Argentina

[7] Facultad de Lenguas y Educación, Universidad Nebrija, Madrid, Spain

[8] Production, Operations and Decision Sciences Area, XLRI, Xavier School of Management, Jamshedpur, India

Abstract

Several studies have examined how music may affect the evaluation of food and drink, but the vast majority have not observed how this interaction unfolds in time. This seems to be quite relevant, since both music and the consumer experience of food/drink are time-varying in nature. In the present study we sought to fix this gap, using Temporal Dominance of Sensations (TDS), a method developed to record the dominant sensory attribute at any given moment in time, to examine the impact of music on the wine taster's perception. More specifically, we assessed how the same red wine might be experienced differently when tasters were exposed to various sonic environments (two pieces of music plus a silent control condition). The results revealed diverse patterns of dominant flavours for each sound condition, with significant differences in flavour dominance in each music condition as compared to the silent control condition. Moreover, musical correspondence analysis revealed that differences in perceived dominance of acidity and bitterness in the wine were correlated in the temporality of the experience, with changes in basic auditory attributes. Potential implications for the role of attention in auditory flavour modification and opportunities for future studies are discussed.

* To whom correspondence should be addressed. E-mail: qianjanice.wang@food.au.dk

Keywords
Crossmodal correspondences, temporal dominance of sensations, attention, wine evaluation, music

1. Introduction

Over the last decade, a rapidly-growing body of empirical research has demonstrated the existence of what is often called 'sonic seasoning', whereby soundtracks with congruent taste/flavour attributes have been shown to influence the perception of what we eat and drink (e.g., Crisinel *et al.*, 2012; Reinoso Carvalho *et al.*, 2017; Wang and Spence, 2016). Crisinel *et al.* (2012) first demonstrated sonic seasoning in a study in which the participants were given samples of bittersweet cinder toffee to evaluate the taste while listening to one of two soundscapes. The soundscapes had been specially composed to match either sweetness or bitterness. Listening to the higher pitched sweet soundscape resulted in the toffee being rated as tasting significantly sweeter and less bitter than while listening to the lower-pitched bitter soundscape instead.

One plausible theory behind sonic seasoning relies on the role of attention; more specifically, the claim is that sound–flavour correspondences may help direct (either automatically, or voluntarily) our attention to certain aspects of the flavours in a food or drink (see Spence *et al.*, 2019, for a review). Attention is intrinsic to how we perceive sensory inputs (Chen *et al.*, 2013), and likely plays a crucial role in determining what we perceive in food (Spence *et al.*, 2000). For instance, Stevenson (2012) illustrated the role of attention in flavour perception in his review by arguing that the reason why we perceive flavour as coming from the mouth, even though smells are captured by the olfactory receptors in the nasal cavity, is due to attentional capture by somatosensory cues (i.e., over olfaction; though see Spence, 2016, for a critical evaluation of the claim). Moreover, when it comes to flavour mixtures, attended flavour elements become relatively more salient than relatively less attended elements (Ashkenazi and Marks, 2004; Marks and Wheeler, 1998; Rabin and Cain, 1989). Crossmodal correspondences between pitch and spatial location have been shown to modulate attentional orienting (e.g., Chiou and Rich, 2012; Klapetek *et al.*, 2012; Mossbridge *et al.*, 2011; Parrot *et al.*, 2015), so it is conceivable that auditory stimuli (specifically taste/flavour-congruent soundtracks) may also be able to shift our attention towards specific tastes/flavours. Moreover, since music and food/drink are both time-varying in nature, it seems only appropriate to take temporality into account when studying the impact of music on the eating/drinking experience.

It should be noted that multisensory congruency has been documented to influence attentional selection in the case of perceptually ambiguous stimuli (van Ee *et al.*, 2009). In their study, van Ee and colleagues demonstrated that congruent auditory or tactile information aided attentional control over competing visual stimuli. Therefore, we might potentially view the phenomenon of 'sonic seasoning' as a specific case of attentional selection, whereby sound–flavour congruence aids attentional selection for congruent flavours present in a mixture (e.g., wine).

In the present study, we used the method of Temporal Dominance of Sensations (TDS). TDS is a relatively recent technique used to record several sensory attributes simultaneously over time. TDS was first introduced at the 5th Pangborn symposium (Pineau *et al.*, 2009) and quickly caught on within the food science community. This technique requires the participant to keep assessing the most dominant flavour attribute, among multiple possible flavour attributes, over the period of assessment. The participant was instructed to assess which flavour attribute is perceived as dominant at any given point in time. TDS has been used to characterise beverages such as blackcurrant squash (Ng *et al.*, 2012) and wine (Meillon *et al.*, 2009; Sokolowsky and Fischer, 2012). Recently, Kantono and his colleagues used TDS to study how liked/disliked music might influence the perception of gelato in terms of basic tastes (Kantono *et al.*, 2016, 2018). We decided to use TDS in order to explore the effect of music on how the taste of wine, a beverage with a complex array of flavours, is perceived. In fact, various multisensory interaction effects on wine flavour perception have been demonstrated with coloured lights (Oberfeld *et al.*, 2009; Spence *et al.*, 2014), tactile stimuli (Wang and Spence, 2018a), music (Spence et al., 2013; Wang and Spence, 2015, 2018b), and combinations of light, soundscapes, and other ambient sensory effects (Velasco *et al.*, 2013). If music does indeed direct the taster's attention to specific flavours, those attended flavours would be more salient/dominant (Ashkenazi and Marks, 2004; Marks and Wheeler, 1998), and different patterns of dominant flavours would be expected under different auditory conditions. Furthermore, we conducted content analysis on the soundtracks used in the study, in order to understand which auditory properties might lead to differences in the temporal patterns of the dominant flavour attributes.

2. Methods

2.1. Participants

A total of 39 (see Note 1) participants took part in the study. Of these, 21 (11 women, 10 men), aged 21–69 years ($M = 37.6$, SD $= 12.8$), participated in the main study and 18 (4 women, 14 men), aged 21–41 years ($M = 27.1$, SD $= 5.5$), took part in a control experiment to assess the hypothesis that the 'musical flavours' dominate over the perceived wine flavours. The participants were recruited at the University Tres de Febrero (UNTREF, Buenos Aires,

Argentina). All the participants gave their informed consent to take part in the study. None of the participants reported a cold or any other known impairment of their sense of smell, taste, or hearing at the time of the study. The study was approved by the Central University Research Ethics Committee of Oxford University (MSD-IDREC-C1-2014-205).

2.2. Auditory Stimuli

Two pieces of music were chosen for the study that varied in tempo, mode, and instrumentation. The first piece was Brian Eno's "Discreet Music", and the second piece was Mussorgsky's "A Night on Bald Mountain". A 45 s excerpt was taken from the beginning of each piece. Note that both of these pieces have been used in various wine–music demos and talks related to sound–taste interactions previously (e.g., Burzynska, 2018). A pre-test ($n = 19$) revealed that both pieces of music led to a significant change in the perceived fruitiness and tannin levels of a light red wine (Georges du Boeuf's Beaujolais-Village, 2014), with the Eno soundtrack enhancing the perceived fruitiness and decreasing the perceived tannin levels, as compared to the Mussorgsky soundtrack. The Eno soundtrack is fairly static throughout, with moderate tempo (70 beats per minute) and consonant harmony, whereas the Mussorgsky soundtrack has dynamic changes in orchestration, register, and loudness; fast tempo (121 beats per minute), and both consonant and dissonant harmonies.

2.3. Wine

The wine used in the study was a Pinot Noir produced in Argentina — Manos Negras Red Soil Select Pinot Noir, 2014. The wine has 13.9% alcohol by volume, 3.92 pH, and 5.70 g/L of Total Acidity (TA), and had been aged for 12 months in 20% new French oak casks (Note 2). A Pinot Noir was used because relatively light-bodied red wines have been commonly used previously in sound–taste demonstrations. The wine was served in 15 mL samples, inside clear plastic 50 mL cups, at room temperature (between 16 and 22°C).

2.4. Design and Procedure

Each participant was seated in front of a computer monitor with headphones and a cup of water to cleanse their palate. The experiment was programmed using the Sensomaker tool for the sensorial characterisation of food products (Nunes and Pinheiro, 2015; Pinheiro *et al.*, 2013).

At the onset of each trial, the participants were given a sample of wine by the experimenter. They were instructed to start the trial as soon as the wine entered their mouth, and to hold it there for the 45 s duration of the trial. The choice of 45 s was informed by other TDS studies (e.g., Kantono *et al.*, 2016, 2019). During this time, the TDS computerised system displayed the entire list of eight adjectives in two columns (red fruit, tannins, alcohol, woody, sweet, acidic, spicy, and bitter) to the participants. The attributes were selected on the basis of a similar TDS study on Pinot Noir wines (Visalli, 2016). The participants were instructed to click on the start button as soon as the wine sample entered their mouth, and then to consider which attribute was perceived as the most dominant. Each time they felt that their perception had changed, they were to click on the new attribute that they perceived to be most dominant. The participants were free to select an attribute several times during the course of the trial, or not at all. The participants first practiced with a weak yerba mate tea solution one to three times to ensure that they could operate the TDS software with ease.

The order in which the adjectives were presented was randomised for each participant in order to avoid any order effect of the list of attributes. However, for a specific participant, the order was always the same and so learning the terms and scoring was facilitated. We used a within-participant full factorial design, with each participant tasting three wine samples in the three auditory conditions, without knowing that the wines were indeed the same. The participants always tasted the wine in the silent condition first, but the order of the two music soundtracks was randomised. In those trials involving a soundtrack, the experimenter started the music at the same time as the participant clicked on the start button. Participants were informed at the start of the soundtrack trials that background music would be presented, but they were not informed about the purpose of the music nor how the music was selected. The entire experiment lasted for around 10 minutes and the participants were debriefed afterwards.

2.5. Data Analysis

TDS curves were produced by the SensoMaker software. They were averaged over all participants and smoothing was applied. Each graph had two additional lines. One, the 'chance level,' is the dominance rate that an attribute would be chosen by chance, in this case equal to 1/8, since there are eight attributes. The second line shows the 'significance level,' which is the minimum value for the dominance level to be considered significantly greater than chance, calculated using the confidence interval of a binomial proportion based on a normal approximation (Pineau *et al.*, 2009). In order to understand specifically how music influenced TDS responses, pairwise correlations between the musical features (frequency content, intensity, and musical segmentation) and the reported dominant tastes were calculated.

3. Results

Figure 1 shows the TDS curves for each of the three auditory conditions. Concentrating on dominance ratings above the significance levels, three major

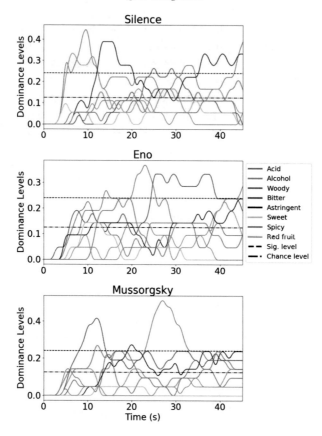

Figure 1. Experiment 1's TDS graphical representation for Manos Negras Pinot Noir wine, in the three auditory conditions. Top: Silent control condition; Middle: Eno soundtrack; Bottom: Mussorgsky soundtrack. The curves were averaged over all participant data and smoothed.

differences can immediately be seen. First, onset time for acidity occurred at around 9 s in the silent baseline condition, whereas acidity peaked at 23 s for the Eno soundtrack and at 27 s for the Mussorgsky soundtrack. Second, bitterness was prominent during 25–38 s for the Eno soundtrack, whereas it was prominent during 8–14 s for the Mussorgsky soundtrack (and was barely registered at 29 s in the silent condition). Finally, in the silent condition, both alcohol and astringency were at significant dominance levels (alcohol between 5 and 10 s, astringency at 10–20 s, then 35–45 s), but was not significant when either of the two soundtracks were played. For the Eno soundtrack, acidity was registered before bitterness, whereas for the Mussorgsky soundtrack, bitterness was registered before acidity.

For each sound condition, the total number of citations (i.e., number of times chosen) as well as the duration of dominance for each of the eight descriptors were calculated. Since none of the measures were normally distributed according to the Shapiro–Wilk test, we used non-parametric rank-sum tests. A Friedman test revealed there to be no significant differences in the total number of adjectives participants' used for each of the three auditory conditions [$\chi^2(2) = 4.00$, $p = 0.14$]. Neither were there any significant differences in dominance durations of acidity [$\chi^2(2) = 2.00$, $p = 0.37$], alcohol [$\chi^2(2) = 2.92$, $p = 0.23$], woodiness [$\chi^2(2) = 3.86$, $p = 0.14$], sweetness [$\chi^2(2) = 1.81$, $p = 0.41$], spiciness [$\chi^2(2) = 2.68$, $p = 0.26$], or red fruit [$\chi^2(2) = 1.27$, $p = 0.53$].

There was, however, a significant difference in dominance durations for bitterness [$\chi^2(2) = 7.75$, $p = 0.021$] and astringency [$\chi^2(2) = 7.55$, $p = 0.023$]. Post-hoc analysis with Wilcoxon signed-rank tests revealed that compared to the silent condition, there were significant increases in bitterness dominance durations for both the Eno soundtrack ($Z = -2.28$, $p = 0.023$) and the Mussorgsky soundtrack ($Z = -2.30$, $p = 0.021$). There were, however, no differences in bitterness between the two soundtrack conditions ($Z = -0.024$, $p = 0.98$).

There were also significant reductions in the durations of astringency dominance for both the Eno soundtrack ($Z = -2.04$, $p = 0.042$) and the Mussorgsky soundtrack ($Z = -2.19$, $p = 0.029$), as compared to the silent condition. There were, once again, no differences between the two soundtrack conditions ($Z = -0.75$, $p = 0.45$).

In order to determine the effect of each soundtrack on the perceived taste of the wine more clearly, TDS difference curves were plotted (see Fig. 2) to reveal the net influence of background music on wine perception, while controlling for how the wine tastes in the silent control condition. To calculate the difference curves, the differences between each soundtrack and the silent condition were plotted at points where they were significantly different from zero by comparing two binomial proportions (Pineau *et al.*, 2009). The effect of the Eno soundtrack, compared to the silent baseline condition, highlights an enhancement of bitterness and a reduction of alcohol in the 0–15 s timeframe, and then a reduction in alcohol around the 30 s mark. The effect of the Mussorgsky soundtrack, compared to the silent condition, is a longer and more prominent enhancement of bitterness during the 0–15 s timeframe along with a reduction in acidity and astringency. There follows an enhancement in acidity around the 25–30 s timeframe.

3.1. Analysis of Individual Musical Features in Relation to Taste

Pairwise Pearson correlations were computed between the TDS curves for the tastes that reach significance in the presence of music (bitterness and acidity, see Fig. 1), and different types of time-varying musical features: 1) acoustic features: frequency content (measured by spectral centroid), and sound

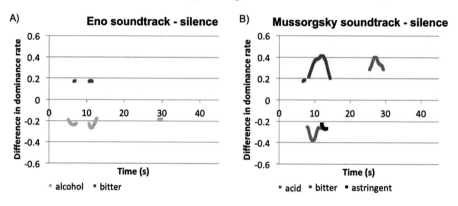

Figure 2. TDS difference curves between (A) the Eno soundtrack and silence condition, and (B) the Mussorgsky soundtrack and silence condition. Only significant differences in dominance ratings between the silent and soundtrack conditions are plotted. In other words, these TDS curves showcase the net influence of background music on wine perception.

intensity (measured by root mean square energy, RMSE); 2) psychoacoustic features: roughness, brightness, inharmonicity; 3) emotion-related features: valence, activity, and tension. These features include many of the major factors in crossmodal taste–sound correspondences that have been documented to date (see Knöferle and Spence, 2012, for a review): spectral centroid (also called spectral balance) is related to timbre brightness/sharpness; energy is related to loudness; roughness, brightness and inharmonicity are timbre features while valence is connected with pleasantness. Another relevant feature, tempo, was almost constant for each musical excerpt. Pitch was not analysed since the music is polyphonic (that is, several pitches are present at the same time). We used MIRToolbox (Lartillot and Toiviainen, 2007) for the computation of the musical curves from the audio files, with an overlapping running window (window length $= 0.05$ s, overlap $= 0.025$ s). If there were to be an influence of a musical feature on taste, we would expect a positive time delay between the music onset time and its effect on the TDS response (due to the time required for auditory processing, choosing an attribute, and finally clicking on it). This delay was estimated as the averaged time $\overline{T}_1$ of first response, corresponding to the time it took the participant to choose a first attribute after the start of the music. We calculated similar average delays for the two music pieces (Mussorgsky: $\overline{T}_{1Muss} = 8.4$ s with SD $\sigma_{Muss} = 4.2$ s, Eno: $\overline{T}_{1Eno} = 8.1$ s with SD $\sigma_{Eno} = 2.6$ s), and for the silent condition ($\overline{T}_{1silence} = 6.9$ s with SD $\sigma_{silence} = 1.9$ s).

In Fig. 3, we plotted the pairwise correlations among the musical parameters and taste attributes for different positive delay windows of the taste curves. We take the relevant lags as those in the intervals ($\overline{T}_{1music} - \sigma_{music}$, $\overline{T}_{1music} + \sigma_{music}$) and ($\overline{T}_{1silence} - \sigma_{silence}$, $\overline{T}_{1silence} + \sigma_{silence}$) marked by the

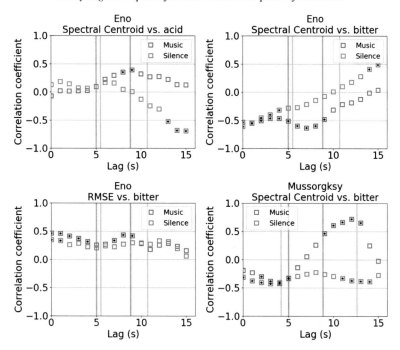

Figure 3. Pearson correlations between musical features and lagged taste curves in the music condition (blue) and silent condition (orange). Stars denote significant correlations. Coloured vertical bars delimit the span of relevant lag times (see text for details).

coloured vertical bars in the figure. The specific pairs of music parameters and taste curves plotted in Fig. 3 were the only ones for which significant correlations were found in the music condition (in blue), while no significant correlation was observed in the silent control condition (in orange) for the relevant delay windows. Moreover, they have important and perceptible variations during the music: the spectral centroid ranged from 946 Hz to 3336 Hz in the Mussorgsky soundtrack, and from 790 Hz to 1091 Hz in the Eno soundtrack; RMSE varied from 0.0006 to 0.09 in the Mussorgsky soundtrack (-45 dB to -3 dB), and from 0.003 to 0.05 in the Eno soundtrack (-34 dB to -11 dB). Another important factor, sensorial dissonance, measured by psychoacoustic roughness (Bigand *et al.*, 1996; Johnson-Laird *et al.*, 2012), gave very similar correlations to those of the RMSE, which are not shown in Fig. 3. Note that even if there were a significant correlation between a music parameter and a taste curve in the silent condition, this would be merely coincidental (since, in fact, people were not listening to any music!). We also considered the significant correlations that appear outside the relevant delay window to be meaningless in this context: since the delay between the curves would be either too short or too long in comparison to the estimated response delay, we do not consider these correlations as representing an influence of music on taste.

In agreement with previous results in the literature (Knöferle and Spence, 2012; Mesz *et al.*, 2011), when the Eno soundtrack was playing in the background, the dominance of acidity was positively correlated with a high spectral centroid, and the dominance of bitterness was correlated with a low spectral centroid. Furthermore, the dominance of bitterness was positively correlated with sound intensity (RMSE) and sensorial dissonance for the Eno soundtrack. On the other hand, for the Mussorgsky soundtrack, a correlation was observed between the dominance of bitterness and a high spectral centroid, contrary to what we observed for the Eno soundtrack.

3.2. Influence of Musical Structure on Taste

We also explored a possible correspondence between the structural segments in the musical excerpts and regions where significant taste evaluations were prominent. A method for novelty-based segmentation was used to help in locating those points in time from a music signal that would correspond to the changes on instrumentation and dynamics (Müller, 2015). Novelty was computed by inspecting the recurrence matrix of the Mel Frequency Cepstral Coefficients (MFCC) audio descriptor and measuring the edges in the block-like structures typically found in the recurrence matrix. This was achieved by convolving an edge detection kernel over the recurrence matrix following the principal diagonal direction. The output of this process was a signal that indicates the novelty as a continuous signal, and the peaks of this novelty signal gave the boundaries of the musical segments (Foote, 2000).

We plotted acidity and bitterness taste curves and overlaid novelty boundaries on top (see Fig. 4). Only for the Mussorgsky soundtrack did we find significant differences between bitterness and acidity, which occur after the

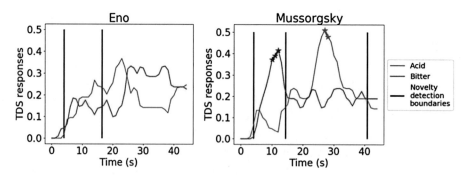

Figure 4. Musical segments and taste curves averaged across participants. Black vertical bars mark novelty peaks in the music. Asterisks mark points of significant difference between bitterness and acidity curves.

boundaries with a delay within one standard deviation of $\overline{T_1}$. Between the first and second boundaries there was significantly more bitterness than acidity, and between the second and third boundaries, there was significantly more dominance in acidity than bitterness. The three boundaries correspond to important points in the music (see video in the Supplementary Materials): the first boundary, at 4.08 s, marks the entrance of excited glissandos in the high register of the woodwind (simultaneously there are chromatic triplets in the violins and a rumbling bass line in the low strings). It is also during this period where bitterness is perceived as significantly more dominant than acidity. The second boundary, at 14.39 s, coincides with the first theme in bassoons, trombones, tuba, violas and low strings. Next, the horns and trumpets enter playing a D, creating a dissonance with the C in the melody. The dissonance continues until it is halted by two loud chords, occurring during the period when perceived acidity was more dominant than bitterness. At the third boundary, the orchestra decays together with the acidity curve, then two more accentuated chords lead to a trill and a general pause (see videos of the music vs the taste curves in the Supplementary Material).

4. Discussion

The results of the present study demonstrate that tasting wine while listening to different soundtracks leads to different perceptions of dominant flavours. Overall, the onset of acidity was earlier in the silent condition than in either of the soundtrack conditions, and astringency was less noticeable when there was music playing. Bitterness was more prominent in the beginning when the wine was tasted when listening to the Mussorgsky soundtrack, whereas for the Brian Eno soundtrack, bitterness in the wine came after the initial registration of acidity. Analysing dominance durations supported results from TDS difference curves, where bitterness was dominant longer — but astringency shorter — during the two soundtrack conditions when compared to the silent condition. Furthermore, while alcohol was dominant in the 10–20 s interval in the silent condition, it was not at a significant dominance level during either of the soundtracks. This implies that music could potentially distract participants from perceiving alcohol accurately, especially when music was presented in combination with the cognitively demanding task (Stafford *et al.*, 2012, 2013).

There were no significant differences in the number of adjectives participants selected — on average the number was around four, or half of the eight available adjectives. For this group of participants, basic tastes such as acidity and bitterness were used more often ($Dur_{acid} = 9.1$ s, $Dur_{bitter} = 7.4$ s) than more descriptive terms such as red fruit and woody ($Dur_{red\_fruit} = 1.6$ s, $Dur_{woody} = 3.7$ s). This might either be attributable to the fact that the participants simply did not taste the more descriptive attributes in the wine, or that

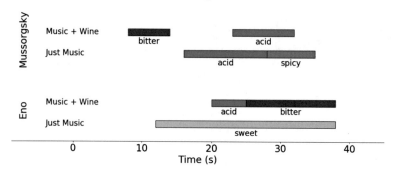

Figure 5. TDS bandplot curves of results with music and wine vs just music, for the Mussorgsky and Eno soundtracks. Only dominant flavours significantly greater than zero are plotted.

basic tastes were simply more dominant (or more easily came to mind) in the wine, especially under experimental conditions.

What are the mechanisms underlying the changes in taste perception in the presence of music? A plausible hypothesis concerning the influence of the music on the flavour in this experiment is that, at least at some moments, the flavour labels are associated with the music, independently of the taste of the wine, and then attributed to the wine. To examine this hypothesis, we performed a control test in which a group of participants ($n = 18$) had to evaluate the music alone, without drinking, using the TDS protocol with the same labels used for the wine. The dominance regions of different musical 'flavours' for this test are shown in Fig. 5. While, in the case of the Eno soundtrack, there is no immediate correspondence between the music-only TDS test and the music + wine TDS test, for Mussorgsky's music, we found an overlap of the acidity-dominant regions in both tests (from 23 s to 27 s), with a delay between them within one standard deviation of $\overline{T_1}$ (see Fig. 5).

A number of possible explanations can be advanced for the overlap of semantic regions for music and taste. On the most superficial level, the participants could have been applying a recognition heuristic (Goldstein and Gigerenzer, 2002). Recognition heuristics consist, in general, in choosing a known alternative over an unknown one; in our case, this implies that having picked a descriptor significant in the music, a participant would prefer to apply this label (e.g., acidity) to an ambiguous, unrecognized taste. However, if the participants merely applied labels from the music to the wine, we would have expected to have seen sweet being chosen for the Eno soundtrack, both when there was music alone or with the combination music and wine. This implies that the recognition heuristic was not the *only* mechanism involved in the participants' TDS ratings.

Besides semantics, more general crossmodal associations between musical features and taste can be relevant for altering wine's flavours. As we hypothesised, specific crossmodal associations can shift people's attention to certain flavours in the wine, which could then make it appear dominant. For instance, acidity/sourness has been reliably associated with a high spectral centroid (Knoeferle *et al.*, 2015; Simner *et al.*, 2010) and correspondingly, a higher spectral centroid in the music was correlated with the dominance of perceived acidity in the Eno soundtrack. Is should be said that some conflicting evidence also appeared for the Mussorgsky soundtrack, where the spectral centroid was positively correlated with bitterness, but according to the literature, bitterness is associated with a low spectral centroid (Knoeferle *et al.*, 2015). However, it is worth noting that this initial section was also highly dissonant, which is correlated in the literature with both bitterness and acidity. Therefore, the Mussorgsky soundtrack, while technically having a high spectral centroid, contained auditory attributes associated with bitterness as well as sourness.

An important theoretical question here regarding the attention hypothesis concerns the temporal resolution of auditory and flavour attention. In the analysis of musical structural correlations to perceived tastes, we took into account the delay time between the participant hearing a feature in a given piece of music, and the participant choosing a specific flavour as dominant. This lag time was approximated by the average time it took the participant to choose an attribute following the onset of the music. For both soundtracks, the lag time was around 8 s (8.1 s for the Eno soundtrack, and 8.4 s for the Mussorgsky soundtrack). For the silent condition, the average lag time was not significantly shorter at 6.9 s. This agrees with the average time before first citation in other TDS studies involving wine (Galmarini *et al.*, 2018; Meillon *et al.*, 2009). That is to say, it would appear to take participants around 7 to 8 s to perceive and select a dominant taste in wine, no matter whether music is present or not (which is understandable as it takes approximately 200 ms for people to register auditory loudness; von Békésy, 1963). Interestingly, this delay time also agrees with the time required before a participant can make emotional judgments about a piece of music, which is also 8 s (Bachorik *et al.*, 2009). Therefore, it is plausible that differences in TDS ratings could also be due to sensation transference (Biggs *et al.*, 2016; Kantono *et al.*, 2019; Wang and Spence, 2018c), where participants could have transferred emotions experienced from listening to the music to the TDS task (for instance, negative feelings from the music could have resulted in ratings of acidity or bitterness). However, the difference in TDS responses between the music-only condition in the control study and the music + wine condition in the main study does not entirely support the sensation transfer hypothesis.

It is worth stressing that the present study has several limitations. First, we had a fairly small sample size ($n = 21$ for the main study, $n = 18$ for the control study), so that any variability in individual taste perception could have altered the pattern of results obtained. Furthermore, we did not control for the levels of wine expertise, although all participants were self-identified wine novices. Wine novices might make less consistent ratings when it comes to wine compared to wine experts (Tempere *et al.*, 2016). Another limitation with the design of the present study was that the silent condition was always presented first, so when participants experienced the soundtrack conditions they had already tasted the wine on one previous occasion (even if they might not have known that it was the same wine). Research has shown that when it comes to tasting several wines in a flight, the order in which the wines are tasted can play a large role in the judgment of the wines (Mantonakis *et al.*, 2009). Therefore, the TDS difference curves between wines tasted while listening to music compared to wines tasted in silence could be due, in part, to the fact that the silent condition always came first. However, it should be stressed that the order of appearance of the two soundtracks was fully randomised, so between-music comparisons are not compromised by any possible order effects.

Moreover, we did not account for the participants' level of musical expertise. The same segment of music might possibly be associated with different tastes depending on the participants' musical expertise. For instance, Wang *et al.* (2015) found that a high-pitched piano piece with dissonant tonality was associated with sweetness by musical novices, but with bitterness by those with musical training. This could be due to the fact that people attend to different elements in the music depending on their level of expertise; for example, novices may tend to focus on timbre, whereas experts may tend to focus on harmony (Wolpert, 1990). Furthermore, Reinoso Carvalho and colleagues (2015) have demonstrated that using participants' individual music–chocolate associations produced more robust crossmodal effects (i.e., modulations in chocolate ratings) compared to the music–chocolate matches designed by the experimenters. Given the evidence for individual differences, it is possible that, in the present study, different participants could have attended to different tastes while listening to the same segment of music.

Looking to the future, the fruitful use of TDS in the present study opens many potential avenues of research, with a focus on the temporal aspects of the music listening experience as well as the tasting experience. As discussed in Spence and Wang (2015), most off-the-shelf music is not ideal for research purposes since stylistic changes often occur and, unless one is careful, there is a real danger in a piece of music corresponding to different tastes/flavours. For instance, both Queen's Bohemian Rhapsody as well as the second movement of Mozart's Piano Sonata No. 12 in F Major (K332) vary between major and minor modes, which would correspond to sweet and sour/bitter tastes, respectively (Knöferle and Spence, 2012). Learning more about the temporal characteristics of such 'sonic seasoning' effects could free researchers from such constraints, as well as enable experience designers to create more fluid and sophisticated experiences which take advantage of the evolving nature of both the listening and the eating/drinking experience.

Supplementary Material

Supplementary material is available online at:
https://brill.figshare.com/s/a4c5f025cceb0781d34a

Notes

1. The number of participants was determined by a convenience sampling. As there was no precedence for music–food TDS analysis at the time of study, it was difficult to run a power analysis. However, the sample sizes used in the studies are in line with typical TDS panel sizes (Meillon *et al.*, 2009; Pineau *et al.*, 2009).

2. http://www.manosnegras.com.ar/images/fichas/Pinot-Noir-Red-Soil-Select-EN.pdf.

References

Ashkenazi, A. and Marks, L. E. (2004). Effect of endogenous attention on detection of weak gustatory and olfactory flavors, *Percept. Psychophys.* **66**, 596–608.

Bachorik, J. P., Bangert, M., Loui, P., Lark, K., Berger, J., Rowe, R. and Schlaug, G. (2009). Emotion in motion: investigating the time-course of emotional judgments of musical stimuli, *Music Percept.* **26**, 355–364.

Bigand, E., Parncutt, R. and Lerdahl, F. (1996). Perception of musical tension in short chord sequences: the influence of harmonic function, sensory dissonance, horizontal motion, and musical training, *Percept. Psychophys.* **58**, 125–141.

Biggs, L., Juravle, G. and Spence, C. (2016). Haptic exploration of plateware alters the perceived texture and taste of food, *Food Qual. Pref.* **50**, 129–134.

Burzynska, J. (2018). Assessing *Oenosthesia*: blending wine and sound, *Int. J. Food Des.* **3**, 83–101.

Chen, K., Zhou, B., Chen, S., He, S. and Zhou, W. (2013). Olfaction spontaneously highlights visual saliency map, *Proc. R. Soc. B Biol. Sci.* **280**, 20131729. DOI:10.1098/rspb.2013.1729.

Chiou, R. and Rich, A. N. (2012). Cross-modality correspondence between pitch and spatial location modulates attentional orienting, *Perception* **41**, 339–353.

Crisinel, A.-S., Cosser, S., King, S., Jones, R., Petrie, J. and Spence, C. (2012). A bittersweet symphony: systematically modulating the taste of food by changing the sonic properties of the soundtrack playing in the background, *Food Qual. Pref.* **24**, 201–204.

Foote, J. (2000). Automatic audio segmentation using a measure of audio novelty, in: *Proc. IEEE Int. Conf. Multimed. Expo*, pp. 452–455.

Galmarini, M. V., Dufau, L., Loiseau, A.-L., Visalli, M. and Schlich, P. (2018). Wine and cheese: two products or one association? A new method for assessing wine-cheese pairing, *Beverages* **4**, 13. DOI:10.3390/beverages4010013.

Goldstein, D. G. and Gigerenzer, G. (2002). Models of ecological rationality: the recognition heuristic, *Psychol. Rev.* **109**, 75–90.

Johnson-Laird, P. N., Kang, O. E. and Leong, Y. C. (2012). On musical dissonance, *Music Percept.* **30**, 19–35.

Kantono, K., Hamid, N., Shepherd, D., Yoo, M. J. Y., Grazioli, G. and Carr, B. T. (2016). Listening to music can influence the hedonic and sensory perceptions of gelati, *Appetite* **100**, 244–255.

Kantono, K., Hamid, N., Shepherd, D., Lin, Y. H. T., Brard, C., Grazioli, G. and Carr, B. T. (2018). The effect of music on gelato perception in different eating contexts, *Food Res. Int.* **113**, 43–56.

Kantono, K., Hamid, N., Shepherd, D., Lin, Y. H. T., Skiredj, S. and Carr, B. T. (2019). Emotional and electrophysiological measures correlate to flavour perception in the presence of music, *Physiol. Behav.* **199**, 154–164.

Klapetek, A., Ngo, M. K. and Spence, C. (2012). Does crossmodal correspondence modulate the facilitatory effect of auditory cues on visual search? *Atten. Percept. Psychophys.* **74**, 1154–1167.

Knoeferle, K. M., Woods, A., Käppler, F. and Spence, C. (2015). That sounds sweet: using cross-modal correspondences to communicate gustatory attributes, *Psychol. Market.* **32**, 107–120.

Knöferle, K. and Spence, C. (2012). Crossmodal correspondences between sounds and tastes, *Psychon. Bull. Rev.* **19**, 992–1006.

Lartillot, O. and Toiviainen, P. (2007). A Matlab toolbox for musical feature extraction from audio, in: *Proceedings of the 10th International Conference on Digital Audio Effects*, pp. 237–244. Bordeaux, France.

Mantonakis, A., Rodero, P., Lesschaeve, I. and Hastie, R. (2009). Order in choice: effects of serial position on preferences, *Psychol. Sci.* **20**, 1309–1312.

Marks, L. E. and Wheeler, M. E. (1998). Attention and the detectability of weak taste stimuli, *Chem. Sens.* **23**, 19–29.

Meillon, S., Urbano, C. and Schlich, P. (2009). Contribution of the Temporal Dominance of Sensations (TDS) method to the sensory description of subtle differences in partially dealcoholized red wines, *Food Qual. Pref.* **20**, 490–499.

Mesz, B., Trevisan, M. A. and Sigman, M. (2011). The taste of music, *Perception* **40**, 209–219.

Mossbridge, J. A., Grabowecky, M. and Suzuki, S. (2011). Changes in auditory frequency guide visual–spatial attention, *Cognition* **121**, 133–139.

Müller, M. (2015). *Fundamentals of Music Processing: Audio, Analysis, Algorithms, Applications*. Springer International Publishing, Cham, Switzerland.

Ng, M., Lawlor, J. B., Chandra, S., Chaya, C., Hewson, L. and Hort, J. (2012). Using quantitative descriptive analysis and temporal dominance of sensations analysis as complementary methods for profiling commercial blackcurrant squashes, *Food Qual. Pref.* **25**, 121–134.

Nunes, C. A. and Pinheiro, A. C. M. (2015). *Sensomaker User Guide*. http://ufla.br/sensomaker/. Retrieved 6 October 2016.

Oberfeld, D., Hecht, H., Allendorf, U. and Wickelmaier, F. (2009). Ambient lighting modifies the flavor of wine, *J. Sens. Stud.* **24**, 797–832.

Parrott, S., Guzman-Martinez, E., Ortega, L., Grabowecky, M., Huntington, M. D. and Suzuki, S. (2015). Direction of auditory pitch-change influences visual search for slope from graphs, *Perception* **44**, 764–778.

Pineau, N., Schlich, P., Cordelle, S., Mathonnière, C., Issanchou, S., Imbert, A., Rogeaux, M., Etiévant, P. and Köster, E. (2009). Temporal Dominance of Sensations: construction of the TDS curves and comparison with time–intensity, *Food Qual. Pref.* **20**, 450–455.

Pinheiro, A. C. M., Nunes, C. A. and Vietoris, V. (2013). Sensomaker: a tool for sensorial characterization of food products, *Cienc. Agrotecnol.* **37**, 199–201.

Rabin, M. D. and Cain, W. S. (1989). Attention and learning in the perception of odor mixtures, in: *Perception of Complex Smells and Tastes*, D. G. Laing, W. S. Cain, R. L. McBride and B. W. Ache (Eds), pp. 173–188. Academic Press, Sydney, Australia.

Reinoso Carvalho, F., Van Ee, R., Rychtarikova, M., Touhafi, A., Steenhaut, K., Persoone, D., Spence, C. and Leman, M. (2015). Does music influence the multisensory tasting experience? *J. Sens. Stud.* **30**, 404–412.

Reinoso Carvalho, F., Wang, Q. (J.), Van Ee, R., Persoone, D. and Spence, C. (2017). "Smooth operator": music modulates the perceived creaminess, sweetness, and bitterness of chocolate, *Appetite* **108**, 383–390.

Simner, J., Cuskley, C. and Kirby, S. (2010). What sound does that taste? Cross-modal mappings across gustation and audition, *Perception* **39**, 553–569.

Sokolowsky, M. and Fischer, U. (2012). Evaluation of bitterness in white wine applying descriptive analysis, time-intensity analysis, and temporal dominance of sensations analysis, *Anal. Chim. Acta* **732**, 46–52.

Spence, C. (2016). Oral referral: on the mislocalization of odours to the mouth, *Food Qual. Pref.* **50**, 117–128.

Spence, C. and Wang, Q. J. (2015). Wine & music (II): can you taste the music? Modulating the experience of wine through music and sound, *Flavour* **4**, 33. DOI:10.1186/s13411-015-0043-z.

Spence, C., Ketenmann, B., Kobal, G. and Mcglone, F. P. (2000). Selective attention to the chemosensory modality, *Percept. Psychophys.* **62**, 1265–1271.

Spence, C., Richards, L., Kjellin, E., Huhnt, A.-M., Daskal, V., Scheybeler, A., Velasco, C. and Deroy, O. (2013). Looking for crossmodal correspondences between classical music and fine wine, *Flavour* **2**, 29. DOI:10.1186/2044-7248-2-29.

Spence, C., Velasco, C. and Knoeferle, K. (2014). A large sample study on the influence of the multisensory environment on the wine drinking experience, *Flavour* **3**, 8. DOI:10.1186/2044-7248-3-8.

Spence, C., Reinoso Carvalho, F., Velasco, C. and Wang, Q. J. (2019). Extrinsic auditory contributions to food perception & consumer behaviour: an interdisciplinary review, *Multisens. Res.* **32**, 275–318. DOI:10.1163/22134808-20191403.

Stafford, L. D., Fernandes, M. and Agobiani, E. (2012). Effects of noise and distraction on alcohol perception, *Food Qual. Pref.* **24**, 218–224.

Stafford, L. D., Agobiani, E. and Fernandes, M. (2013). Perception of alcohol strength impaired by low and high volume distraction, *Food Qual. Pref.* **28**, 470–474.

Stevenson, R. J. (2012). The role of attention in flavour perception, *Flavour* **1**, 2. DOI:10.1186/2044-7248-1-2.

Tempere, S., Hamtat, M.-L., de Revel, G. and Sicard, G. (2016). Comparison of the ability of wine experts and novices to identify odorant signals: a new insight in wine expertise, *Aust. J. Grape Wine Res.* **22**, 190–196.

van Ee, R., van Boxtel, J. J. A., Parker, A. L. and Alais, D. (2009). Multisensory congruency as a mechanism for attentional control over perceptual selection, *J. Neurosci.* **29**, 11641–11649.

Velasco, C., Jones, R., King, S. and Spence, C. (2013). Assessing the influence of the multi-sensory environment on the whisky drinking experience, *Flavour* **2**, 23. DOI:10.1186/2044-7248-2-23.

Visalli, M. (2016). *Results of the Analysis of the Wine-Cheese Data Collected During the Social Program, Presentation Given at EuroSense 2016*. Dijon, France.

von Békésy, G. (1963). Hearing theories and complex sounds, *J. Acoust. Soc. Am.* **35**, 588–601.

Wang, Q. (J.) and Spence, C. (2015). Assessing the effect of musical congruency on wine tasting in a live performance setting, *i-Perception* **6**. DOI:10.1177/2041669515593027.

Wang, Q. (J.) and Spence, C. (2016). 'Striking a sour note': assessing the influence of consonant and dissonant music on taste perception, *Multisens. Res.* **29**, 195–208.

Wang, Q. J. and Spence, C. (2018a). A smooth wine? Haptic influences on wine evaluation, *Int. J. Gastron. Food Sci.* **14**, 9–13.

Wang, Q. (J.) and Spence, C. (2018b). Assessing the influence of music on wine perception amongst wine professionals, *Food Sci. Nutr.* **52**, 211–217.

Wang, Q. (J.) and Spence, C. (2018c). 'A sweet smile': more than merely a metaphor? *Cogn. Emot.* **32**, 1052–1061.

Wang, Q. (J.), Woods, A. and Spence, C. (2015). "What's your taste in music?" a comparison of the effectiveness of various soundscapes in evoking specific tastes, *i-Perception* **6**. DOI:10.1177/2041669515622001.

Wolpert, R. S. (1990). Recognition of melody, harmonic accompaniment, and instrumentation: musicians vs. nonmusicians, *Music Percept.* **8**, 95–105.

Index

Printed in the United States
By Bookmasters